SpringerBriefs in History of Science and Technology

Series Editors

Gerard Alberts, University of Amsterdam, Amsterdam, The Netherlands

Theodore Arabatzis, University of Athens, Athens, Greece

Bretislav Friedrich, Fritz Haber Institut der Max Planck Gesellschaft, Berlin, Germany

Ulf Hashagen, Deutsches Museum, Munich, Germany

Dieter Hoffmann, Max-Planck-Institute for the History of Science, Berlin, Germany

Simon Mitton, University of Cambridge, Cambridge, UK

David Pantalony, Ingenium - Canada's Museums of Science and Innovation/ University of Ottawa, Ottawa, ON, Canada

Matteo Valleriani, Max-Planck-Institute for the History of Science, Berlin, Germany

The *SpringerBriefs in the History of Science and Technology* series addresses, in the broadest sense, the history of man's empirical and theoretical understanding of Nature and Technology, and the processes and people involved in acquiring this understanding. The series provides a forum for shorter works that escape the traditional book model. SpringerBriefs are typically between 50 and 125 pages in length (max. ca. 50.000 words); between the limit of a journal review article and a conventional book.

Authored by science and technology historians and scientists across physics, chemistry, biology, medicine, mathematics, astronomy, technology and related disciplines, the volumes will comprise:

1. Accounts of the development of scientific ideas at any pertinent stage in history: from the earliest observations of Babylonian Astronomers, through the abstract and practical advances of Classical Antiquity, the scientific revolution of the Age of Reason, to the fast-moving progress seen in modern R&D;
2. Biographies, full or partial, of key thinkers and science and technology pioneers;
3. Historical documents such as letters, manuscripts, or reports, together with annotation and analysis;
4. Works addressing social aspects of science and technology history (the role of institutes and societies, the interaction of science and politics, historical and political epistemology);
5. Works in the emerging field of computational history.

The series is aimed at a wide audience of academic scientists and historians, but many of the volumes will also appeal to general readers interested in the evolution of scientific ideas, in the relation between science and technology, and in the role technology shaped our world.

All proposals will be considered.

Andrea Strazzoni

The Manuscript Dissemination of Descartes's *Traité de l'homme*

With an Edition of the *Tractatus de homine a Cartesio*

 Springer

Andrea Strazzoni
Università degli Studi di Torino
Dipartimento di Studi Storici
Turin, Italy

ISSN 2211-4564 ISSN 2211-4572 (electronic)
SpringerBriefs in History of Science and Technology
ISBN 978-3-031-72374-2 ISBN 978-3-031-72375-9 (eBook)
https://doi.org/10.1007/978-3-031-72375-9

© The Author(s), under exclusive license to Springer Nature Switzerland AG 2025

This work is subject to copyright. All rights are solely and exclusively licensed by the Publisher, whether the whole or part of the material is concerned, specifically the rights of translation, reprinting, reuse of illustrations, recitation, broadcasting, reproduction on microfilms or in any other physical way, and transmission or information storage and retrieval, electronic adaptation, computer software, or by similar or dissimilar methodology now known or hereafter developed.
The use of general descriptive names, registered names, trademarks, service marks, etc. in this publication does not imply, even in the absence of a specific statement, that such names are exempt from the relevant protective laws and regulations and therefore free for general use.
The publisher, the authors and the editors are safe to assume that the advice and information in this book are believed to be true and accurate at the date of publication. Neither the publisher nor the authors or the editors give a warranty, expressed or implied, with respect to the material contained herein or for any errors or omissions that may have been made. The publisher remains neutral with regard to jurisdictional claims in published maps and institutional affiliations.

This Springer imprint is published by the registered company Springer Nature Switzerland AG
The registered company address is: Gewerbestrasse 11, 6330 Cham, Switzerland

If disposing of this product, please recycle the paper.

Acknowledgments

This publication is part of the *NODES* project which has received funding from the MUR—M4C2 1.5 of PNRR funded by the European Union—NextGenerationEU (Grant agreement no. ECS00000036—CUP: D17G22000150001). Previously, the research leading to this publication has received funding from the European Union's Horizon 2020 research and innovation programme under the Marie Skłodowska-Curie grant agreement no. 892794 (*READESCARTES*), and the Swiss National Science Foundation—SNF, Spark grant number CRSK-1_190670 (*Testing a Multi-Disciplinary Approach to an Unexplored Body of Literature: The Case of Cartesian Dictations*). Special thanks go to the Dipartimento di Studi Storici (Turin), to Harold J. Cook, Elena Corniolo, Marta Gravela, Lisa Kuitert, Gideon Manning, Antonio Olivieri, Rudolf Rasch, Phillip R. Sloan, to the anonymous reviewers who have commented upon the different versions of this work—originating from a panel talk given at the History of Science Society Annual Meeting, Utrecht, 2019—to Jaap de Vreugd and Els Koeneman of the Leiden Universiteitsbibliotheek, to Mark Rogers, and to the Forschungszentrum Gotha der Universität Erfurt.

About This Book

In this book I provide the study of a foremost case of manuscript dissemination and fostering of knowledge production in the early modern age, namely of the covert circulation of a pivotal text in the history of science: the *Traité de l'homme* of René Descartes, completed in 1633, from the 1640s up to the appearance of its first editions in the 1660s. By disclosing unexplored sources and figures, I discuss a number of manuscripts (all but one now lost), shedding light, in particular, on the role of Elisabeth of Bohemia, Henricus Regius, Johannes de Raey and Aernout Huyberts in the dissemination and use of the treatise, and the attempts at securing and publishing it by the Elzeviers, Pierre-Hector Chanut, and Claude Clerselier. On the basis of this reconstruction I provide a study and edition—with an apparatus of different readings—of the handwritten translation of the treatise recently brought to the attention of scholars (ms. ATH 1444), relating it to the efforts at making sense of the text by the Leiden scholars, and showing that it conveys a version of the treatise akin to the one published (1662) by Florentius Schuyl, while Clerselier might have published a text revised by Descartes around 1647.

Contents

1	**Introduction**	1
	1.1 Overview	1
	1.2 The First Autograph and Three Copies	6
	1.3 Regius	10
	1.4 Other Dutch Copies	14
	1.4.1 Heereboord	14
	1.4.2 De Raey	15
	1.4.3 Huyberts	22
	1.5 Clerselier, Chevreau, and Andreae	26
	1.6 The Elzeviers' Editorial Plans	35
	1.7 ATH 1444	38
	1.8 Conclusion	45
2	**Edition of ATH 1444**	47
	Appendix	111
	Bibliography	119
	Index of Names	127
	Index of Manuscripts	131

About the Author

Andrea Strazzoni, Ph.D., (Erasmus University Rotterdam, 2015), is an assistant professor of the history of science at the University of Turin, and Marie-Curie alumnus of the Ca' Foscari University of Venice. He has published monographs and articles on the history of early modern philosophy and science, as well as editions of primary sources, exploring the inclusion of a reflection on scientific practice in the philosophical discourse, the genetic role of teaching practice and handwritten sources in the development of new ideas, and the medical and natural-historical underpinnings of Cartesian science. He is the author of three monographs—*The Quarrel over Swammerdam's Posthumous Works* (Brill, 2023), *Burchard de Volder and the Age of the Scientific Revolution* (Springer, 2019), *Dutch Cartesianism and the Birth of Philosophy of Science* (De Gruyter, 2018)—and of several articles in specialized journals.

List of Figures

Fig. A.1 Schuyl's representation of the constitution of the brain, nerves, and muscles (*Source* Descartes 1662, 20. University of California, Biomed History and Special Collections, signature: WZ 250 D453d 1662. Public domain) 111

Fig. A.2 Van Gutschoven's representation of the nerve-muscle system (*Source* Descartes 1664b, 16. University of California, Biomed History and Special Collections, signature: WZ 250 D453h 1664. Public domain) 112

Fig. A.3 Descartes's representation of the nerve-muscle system (in Clerselier's edition) (*Source* Descartes 1664b, 17. University of California, Biomed History and Special Collections, signature: WZ 250 D453h 1664. Public domain) ... 112

Fig. A.4 La Forge's representation of the nerve-muscle system (*Source* Descartes 1664b, 18. University of California, Biomed History and Special Collections, signature: WZ 250 D453h 1664. Public domain) 113

Fig. A.5 Descartes's representation of the nerve-muscle system (in Schuyl's edition) (*Source* Descartes 1662, 25. University of California, Biomed History and Special Collections, signature: WZ 250 D453d 1662. Public domain) 113

Fig. A.6 Descartes's representation of sounds (in Schuyl's edition) (*Source* Descartes 1662, 43. University of California, Biomed History and Special Collections, signature: WZ 250 D453d 1662. Public domain) 114

Fig. A.7 The representation of sounds in Clerselier's edition (*Source* Descartes 1664b, 36. University of California, Biomed History and Special Collections, signature: WZ 250 D453h 1664. Public domain) 114

List of Tables

Table 1.1   An overview of the manuscripts of Descartes's *L'homme* 3
Table A.1   Textual reliance of Huyberts on Descartes's *L'homme* 115
Table A.2   Paragraph division of ATH 1444 117

Chapter 1
Introduction

Abstract In this chapter I provide a hypothetical reconstruction of the dissemination of the handwritten copies of René Descartes's *Traité de l'homme* up to the appearance of its first printed editions in 1662 (Latin) and 1664 (French). I discuss the manuscripts copied by Antonie Studler van Surck, Alphonse Pollot, and Elisabeth of Bohemia, its dissemination through the Socinian circles (including Henricus Regius's access to it), and among the Leiden Cartesians (Adriaan Heereboord, Johannes de Raey, and Aernout Huyberts). Moreover, I relate the manuscripts to the various attempts at securing and publishing manuscripts of *L'homme* (by the Elzeviers, Pierre-Hector Chanut, and Claude Clerselier). On the ground of this reconstruction I provide a discussion of the handwritten translation of the treatise recently brought to the attention of scholars (ms. ATH 1444), relating it to the efforts at making sense of the text by the Leiden Cartesians, and showing that it conveys a version of the treatise akin to the one published (1662) by Florentius Schuyl (with which it agrees in a number of substantial readings), while Clerselier might have published a text revised by Descartes around 1647.

Keywords René Descartes · *Traité de l'homme* · Early modern handwritten sources · Elisabeth of Bohemia · Henricus Regius · Johannes de Raey · Aernout Huyberts · Elzeviers · Florentius Schuyl · Claude Clerselier · Ms. ATH 1444

1.1 Overview

In recent years the reception of the ideas of René Descartes (1596–1650) has been boosted by a renewed interest for their implications in neurophysiology, the life sciences, and medicine, today widely recognized as a foremost field of interest of his.[1] Accordingly, special attention is being increasingly paid to Descartes's foremost treatise in physiology, namely his *Traité de l'homme*, which he wrote in the early 1630s but which was published only posthumously, in a Latin translation (*De homine*) published in 1662 and 1664 by Florentius Schuyl (1619–1669), and then in

---

[1] Des Chene 2001; Aucante 2006; Caps 2010; Nachtomy and Smith 2014; Ragland 2022.

French by Claude Clerselier (1614–1684) in 1664 and 1677. The treatise had, meanwhile, a wide manuscript circulation: this caused, from its first dissemination in the 1640s, polemics on the use of Descartes's philosophy (as in the case of Henricus Regius), and exchanges between Descartes and Elisabeth of Bohemia. It was, in fact, the only comprehensive text by Descartes in physiology, representing a completion of his philosophical programme and being eagerly awaited by his followers, who in a few cases—such as Regius and Cornelis van Hogelande—published treatises in physiology and medicine inspired by his ideas or by his very *L'homme*, ideally fulfilling and making public Descartes's project, before the appearance of the 1662 and 1664 editions. Moreover, owners of a handwritten copy, such as Adriaan Heereboord, Johannes de Raey, and Tobias Andreae were influential teachers of Cartesian ideas before those editions: so that the manuscript circulation of the treatise potentially represents a key factor—still to be fully explored—in shaping the dissemination of Cartesianism in the 1640s–1660s. Moreover, a reconstruction of this circulation can shed light on the possible, different redactions, by Descartes, of the treatise (as the editions of Schuyl and Clerselier reveal notable different readings, which I will henceforth refer to as 'variants'), and so on his very intellectual path, and offer the basis for an assessment of the differences between its surviving versions: including a recently resurfaced handwritten one (edited in this book), as well as those which appeared after Schuyl's and Clerselier's editions (the exploration of which nonetheless exceeds the scope of the present book).

Notwithstanding the increasing attention to Descartes's *L'homme* and its reception,[2] however, the reconstruction of the manuscript dissemination of this treatise is still limited, with some insights provided by Matthijs van Otegem and Franco A. Meschini.[3] In order to shed more light on the issue, in this book I offer hypotheses on the history of the different manuscripts of Descartes's *L'homme*—including the only extant one (henceforth 'ATH 1444'), viz. a Latin translation of the treatise brought to the attention of the scholarly community by Erik-Jan Bos in 2022[4]—and how their circulation related to the editions by Schuyl and Clerselier, as well as to the editorial projects of others. Given the fact that only one manuscript is currently known, I do not offer a stemmata diagram, but only more circumscribed inferences on the possible filiation of the copies of the treatise, and on their contexts of production

---

[2] In 2016 Delphine Antoine-Mahut and Stephen Gaukroger edited a volume, *Descartes' Treatise on Man and Its Reception*, providing the first systematic exploration of the development, editorial history and reception of Descartes's treatise: Antoine-Mahut and Gaukroger 2016.

[3] Van Otegem 2002, chapter 9; Meschini 2011; Meschini 2015; Meschini 2016. See also Matton 2005.

[4] See the Leiden Special Collections Blog, URL = <https://www.leidenspecialcollectionsblog.nl/articles/an-unknown-latin-manuscript-translation-of-descartes-lhomme>, and URL = <https://www.universiteitleiden.nl/in-de-media/2022/08/onbekende-vertaling-van-rene-descartes-lhomme-ontdekt-in-leidse-bibliotheca-thysiana> (Bos 2022). Accessed 23 March 2024. The manuscript is held at the Bibliotheca Thysiana at Leiden (now part of the Leiden Universiteitsbibliotheek), with signature ATH 1444; catalogued in Van Roijen et al. 2013, 77. See *infra*, n. 152. All internal references are to footnotes in this chapter of the book.

1.1 Overview

Table 1.1 An overview of the manuscripts of Descartes's *L'homme*

Manuscript	Owners	Kind and source	Years of writing
Descartes	René Descartes, Antonie Studler van Surck (on loan; later inherited), Pierre-Hector Chanut, Claude Clerselier	Autograph, original; on loan to Van Surck for copying in 1641–1642; inherited by him around Descartes's death (1650); got by Chanut in mid-1653–late 1655; given to Clerselier during or after July 1654 or around 1656	July 1633 (completion of a first version), around 1647 (revision)
VanSurck	Van Surck; Florentius Schuyl (loaned or gifted)	Copy of *Descartes*, made by Van Surck; loaned or gifted to Schuyl in 1659–1661 in order to allow him to complete his edition, in his possession or accessible to him in late 1663–early 1664	1641–1642
Pollot	Alphonse Pollot	Copy of *VanSurck*, made by Pollot; requested at first (November 1643) from Descartes through Constantijn Huygens	1643–1646
Elisabeth	Elisabeth of Bohemia	Copy of *VanSurck*, made by Elisabeth	July 1644–October 1645
LaVoyette	Louis de La Voyette	Copy of *Elisabeth*, got from her before or during July 1646	Before or during July 1646
Regius	Henricus Regius	Copy of *Elisabeth*, got through Samson Johnson in May–July 1646	May–July 1646
Heereboord	Adriaan Heereboord, Schuyl (on loan; kept after Heereboord's death in 1661)	Copy deriving from *Regius*, got from Regius or Johannes de Raey (or source of the latter's copy); requested at first (April 1642) through Andreas Colvius; loaned to Schuyl in 1659–1661 (slightly before his access to *VanSurck*), at his disposal in late 1663–early 1664	1646–1661
DeRaey	Johannes de Raey	Copy deriving from *Regius*, got from Regius or Heereboord in 1652–1656 (or source of the latter's copy); used for a Latin translation during or after 1664	1652–1656

(continued)

Table 1.1 (continued)

Manuscript	Owners	Kind and source	Years of writing
Huyberts1	Aernout Huyberts	Copy deriving from *Regius*, got from De Raey or Heereboord in 1652–1656	1652–1656
Chevreau	Urbain Chevreau; Chanut (on loan)	Copy of *LaVoyette*, got from him in early 1653; loaned to Chanut for copying in May–August 1653	Early 1653
Chanut	Chanut, Clerselier	Copy of *Chevreau*, commissioned by Chanut at Hamburg; given to Clerselier during or after July 1654	May–August 1653
Andreae	Tobias Andreae	Originating from the circle of the Dutch Cartesians	Before or during 1654
Clerselier2	Clerselier	Copy of *Andreae*; requested by Clerselier January–July 1654, and delivered to him in July–December 1654	July–December 1654
Huyberts2	Huyberts	Copy of *Descartes* (revised version), sent by Clerselier to Huyberts after a request by Louis and Daniel Elzevier in 1657	1657
Schuyl	Schuyl	Copy of *Pollot*, got from Pollot before or in early 1659, as the first text used for Schuyl's edition	Before or in early 1659
VanGutschoven	Gerard van Gutschoven	Copy of *Descartes* (revised version), delivered to Van Gutschoven for the preparation of the figures of Clerselier's edition after a lengthy process in summer 1660	Late 1659–early 1660
LaForge	Louis de La Forge	Copy of *Descartes* (revised version), delivered to La Forge for the preparation of the figures of Clerselier's edition	Summer 1660
Philippi	Guillaume Philippi	Copy of *VanGutschoven*, got from Van Gutschoven in 1660–1661	1660–1661

(continued)

1.1 Overview

Table 1.1 (continued)

Manuscript	Owners	Kind and source	Years of writing
Clerselier1	Clerselier	Copy of one of *Schuyl*, *VanSurck*, or *Heereboord*, or of a collation of (some of) them, requested and got from Schuyl by Clerselier in 1660–1661	1660–1661
ATH 1444	Johannes Thysius (uncertain)	Probably originating from the circle of the Leiden Cartesians	1648–1653, or probably before Schuyl's edition

and dissemination. In order to provide the reader with some guidance, their discussion is preceded by a summary, in the form of Table 1.1, of their owners, kinds and sources, and years of writing (by which they are ordered, an order not necessarily followed in the rest of the book)—which are, generally speaking, for the most part hypothetical (so that, unlike in the rest of the book, I do not use cautious or dubitative formulas in the table, except in the case of ATH 1444). In Sect. 1.2, I discuss the first dissemination of the treatise by an autograph, a copy of it, and two copies made after this copy—belonging around 1641–1646 to Antonie Studler van Surck, Alphonse Pollot, and Elisabeth. Then (Sect. 1.3) I discuss the copy of Regius, who notoriously had access to the text without Descartes's authorization and plagiarized it in 1646: I discuss the hypothesis that he got a copy from Elisabeth through Samson Johnson. In Sect. 1.4 I relate Regius's copy with others circulating among Dutch Cartesians: namely those of Heereboord, De Raey and Aernout Huyberts. I focus especially on De Raey's attack on Schuyl's 1662 edition and on Huybert's efforts to make sense of the text through anatomical observations. Then (Sect. 1.5) I scrutinize how Clerselier could have got the autograph he used for his edition in the light of other copies possibly used as sources for his work: namely those of Urbain Chevreau and Andreae. In the rest of the book, I discuss the Elzeviers' aborted editorial project on the treatise (Sect. 1.6), relating it to the activities of the Dutch Cartesians, and I make some hypotheses and comments on ATH 1444 (Sect. 1.7), of which I offer, after some conclusions (Sect. 1.8), an edition (Chap. 2), with an apparatus of variants taking into account Schuyl's and Clerselier's editions. The book is thus aimed at those interested in the dissemination of Descartes's ideas in physiology, those curious about textual evidence pertaining to the tradition of his works, and at the scholars of the history of the book.

1.2 The First Autograph and Three Copies

Descartes: the first manuscript of Descartes's *L'homme*, was an autograph described by Descartes in his letter to Marin Mersenne (1588–1648) of 23 November 1646 as "so messy that I myself would have great difficulty reading it."[5]

VanSurck: a copy of *Descartes* was probably made by and belonged to Descartes's friend and banker Antonie Studler van Surck (c. 1608–1666). As Schuyl reports in the preface to his edition of the treatise (1662), indeed, to complete his translation he used a copy transcribed by Van Surck himself "as accurately as possible" from an autograph of Descartes, which he "gave" ("concessit"—it being unclear if it was a loan or a gift) Schuyl in order to allow him to complete his edition: this concession probably happened in 1659–1661,[6] with the copy apparently in Schuyl's possession or accessible to him in late 1663–early 1664, as discussed in Sect. 1.4.1.[7] Given the fact that in the already mentioned letter of Descartes to Mersenne of 23 November 1646 Descartes reports that from his autograph (*Descartes*) he allowed an intimate friend of his to make a copy (apparently with no copyist involved), we may suppose that this copy was *VanSurck*. This copy was made with Descartes's permission "4 or 5 years" before the time of the letter, i.e. around 1641–1642.[8]

---

[5] "[I]l y a desia 12 ou 13 ans que i'avois descrit toutes les fonctions du corps humain, ou de l'animal, mais le papier ou ie les ay mises est si brouillé que i'aurois moy mesme beaucoup de peine a le lire; toutefois ie ne pûs m'empescher il y a 4 ou 5 ans de le prester a un intime ami, lequel en fit une copie, laquelle a encore esté transcrite depuis par deux autres auec ma permission, mais sans que ie les aye releuës ny corrigées. Et ie les auois priez de ne le faire voir a personne, comme aussy ie ne l'ay iamais voulu faire voir a Regius," Descartes to Mersenne, 23 November 1646, in AT IV, 566–567. The writing of Descartes's *L'homme* was finished in 1633: see, for instance, Descartes to Mersenne, 22 July 1633, in AT I, 266–269. Unless taken from an English edition, all the translations from primary sources are mine.

[6] See *infra*, n. 104.

[7] "Ad hanc igitur me primum movit (…) Alphonsus Palotti (…), facta mihi copia manuscripti, quod ipse sophiae studiosissimus quam nitidissime descripserat. Additis duabus figuris a Des Cartes rudi Minerva exaratis, quae pag. 25 et 43 referuntur. Pudori meo deinde succurrit et ad opusculum absolvendum atque in lucem edendum impulit authoritas viri, inclyti generis et exquisitissimae doctrinae nobilitate nulli secundus, Anthonius Stutler van Surck (…) qui nativa sua benevolentia ectypum a sese ex authoris nostri autographo quam accuratissime delineatum in hunc finem mihi lubens concessit. Promovit denique et ursit negotium Nobilissimus D. Claudius Clerselier," Descartes 1662, *Ad lectorem*, 32 (unnumbered). Schuyl elsewhere used the verb 'commodo' (to lend) for another manuscript, so we can hypothesize that Van Surck gave his copy as a gift to Schuyl: see *infra*, n. 55. In the course of the book, I use single quotation marks '' whenever I mean a word as such (regardless of its being singular, plural, or of its declension) or I take a word in a peculiar sense. I use double quotation marks "" whenever I quote from a source, even if the quotation concerns words meaning words as such. In the case of my mentioning words meaning words which can be found in a different declension, with respect to my mentioning, in the source referred to for the sake of my argument, or in the case I mention words as such tracing to hypothetical texts, e.g. (1) words supposedly translated in an extant source, but whose original is just presumed, or (2) words which are hypothetical alternative readings of certain given words, I use single quote marks.

[8] See *supra*, n. 5.

1.2 The First Autograph and Three Copies

A supposition which—it should be borne in mind—is based on an interpretation of Schuyl's preface according to which the copy made by Van Surck was not specifically made for Schuyl from an autograph around 1659–1661, but was already in Van Surck's possession since 1641–1642. Of course, it might be that Van Surck made two copies from an autograph: one in 1641–1642 (mentioned by Descartes) and another in 1659–1661 (mentioned by Schuyl). In this case, however, we should suppose the existence of two autographs: *Descartes* (in possession of Van Surck) and another one, in possession of Clerselier at the end of the 1650s and labelled by him as imperfect (as I discuss in Sect. 1.5). This is somehow consistent with the presence of variants between ATH 1444 and the editions of Schuyl on the one hand (whose version(s) seems to trace to one redaction of a text, albeit with some variants between each other, as I will discuss), and those of Clerselier on the other (who seems to have edited a text at some point revised by Descartes, with respect to the one(s) of ATH 1444 and Schuyl). In the rest of this book, though, I will assume that Van Surck provided Schuyl with a copy he had made in 1641–1642, not in 1659–1661: as this (1) is consistent with the role of Van Surck as one of the closest contacts of Descartes in the Netherlands, who might well have copied the treatise already in the early 1640s, and (2) presupposes the existence of just one autograph of *L'homme*: a treatise the re-writing of which in a second autograph would have probably led to a better form than that mentioned by Clerselier, who nonetheless worked on a re-touched version of it (this being consistent with Descartes's own words on his researches in physiology),[9] and probably to one with figures.

Pollot: from the copy of *Descartes* (i.e. *VanSurck*) two other copies were drawn, apparently transcribed by their very owners and neither read nor corrected by Descartes, as declared by him in the same letter to Mersenne, where it is also specified that no further copies were authorized by him, and not even their owners were authorized to show them to anybody.[10] One of the two copies probably belonged to Alphonse Pollot (c. 1602–1668). In his above-mentioned 1662 preface, indeed, Schuyl also reports to have used a copy (*Schuyl*) provided to him by Pollot (most probably before or in early 1659),[11] drawn from a manuscript (i.e. *Pollot*), which Pollot himself had transcribed "as clearly as possible" from a further manuscript (hypothetically, *VanSurck*) on which Schuyl does not provide insights.[12] In fact, Pollot was not the only owner of a manuscript of Descartes's *L'homme* before its first publication in 1662, and there is no direct evidence that he was the owner of a copy drawn from *VanSurck*. However, we can presume that he was the owner of such a copy for the following reasons: (1) Elisabeth of Bohemia (1618–1680) read Descartes's *L'homme* before October 1645 and was the owner of a copy (*Elisabeth*, which I discuss next), while Pollot acted, together with Van Surck, as her intermediary with Descartes,[13]

---

[9] See *infra*, n. 116.
[10] See *supra*, n. 5.
[11] See *infra*, n. 104.
[12] See *supra*, n. 7.
[13] See, for instance, Elisabeth to Descartes, 6 May 1643, in AT III 660–662; Elisabeth to Descartes, 1 July 1643, in AT IV, 1–3.

so it is probable that *VanSurck*, *Pollot* and *Elisabeth* were directly related to each other and all authorized by Descartes. (2) In Schuyl's *Epistola ad amicum* opening his second edition (1664) of Descartes's *L'homme*, both *VanSurck* and *Schuyl*—or *Pollot*, from which the latter was drawn—are labelled as "the most accurate."[14] In turn, another copy he had at his disposal at the latest from June 1661 (and probably still in his hands at the time of the *Epistola*, dated 6 March 1664), when he was presumably still working on his edition (which appeared in summer 1662),[15] namely a copy once belonging to Adriaan Heereboord (1613–1661; *Heereboord*, which I discuss in Sect. 1.4.1), is not mentioned in his 1662 preface, and is not described as to its accuracy in the 1664 *Epistola*. Accordingly, we can presume that *Pollot* was closer to *VanSurck* than to *Heereboord* in quality, i.e. that it was drawn from the latter. In his 1662 preface, in fact, Schuyl claims to have used *VanSurck* to finish his work—so that probably he was looking for a manuscript clearer than *Schuyl* (or *Pollot*) in order to make sense of the text, and having had access to *Heereboord*, found it unreliable and reverted in the same period—around 1659–1661—to *VanSurck* (regardless of the intrinsic difficulties in understanding the contents of a figureless text, as I discuss in a moment). (3) In his 1662 preface Schuyl mentions to have received from Pollot also two autograph figures by Descartes, which he used for his edition: namely a figure representing nerves and muscles and a diagram consisting of lines divided into parts and representing sounds,[16] both of which I discuss in Sect. 1.5. Therefore, we can suppose that Pollot could have personally made his copy from the one of Van Surck, who in turn had direct access to *Descartes* and, presumably, related materials like the two autograph figures. These were not copied together with *Descartes*, which was figureless, as Descartes knew that its copies were figureless and Schuyl and Clerselier had, notoriously, to supply them.[17] Alternatively, Pollot got such autograph figures directly from Descartes (keeping them after his death), and a fortiori he was authorized by him to own a copy.

Elisabeth: the other copy of *VanSurck* probably belonged to Elisabeth. Descartes himself, indeed, reveals, through his letters to her of 6 October 1645 and 31 January 1648, that she had seen and read his *L'homme*,[18] while more direct evidence of her being in possession of a copy (which we can suppose was authorized by Descartes, as he was well aware she had read the treatise) is provided in a *memoire* of Urbain Chevreau (1613–1701).[19] According to Chevreau, when Pierre-Hector Chanut (1601–1662) came to Sweden for the last time (around May 1653, staying there for about a month),[20] Chevreau communicated to him that he, Chevreau, had

---

[14] The relevant text of Schuyl's *Epistola* is quoted *infra*, n. 55.

[15] See Niels Steensen to Thomas Bartholin, 26 August 1662, in Bartholin 1663–1667, volume 3, 103–113.

[16] See *supra*, n. 7.

[17] See *infra*, n. 26.

[18] Descartes to Elisabeth, 6 October 1645, in AT IV, 309–310; Descartes to Elisabeth, 31 January 1648, in AT V, 112.

[19] Discussed in Matton 2005.

[20] As reconstructed in Sect. 1.5, where the text of the *memoire* is provided.

a copy of Descartes's *L'homme*, a treatise which Chanut had looked for over a long time. Chevreau reports to have obtained it from Louis de La Voyette (d. 1659)— in early 1653, when they both were at Stockholm—and who in turn had it from Elisabeth, probably before she moved to Germany in summer 1646 (as I discuss in Sect. 1.3), it being in any case unclear whether it was the same copy (*Elisabeth*), or if intermediate copies (*LaVoyette, Chevreau*) had been made. It was from the copy lent to Chanut by Chevreau, according to the latter, that Chanut made a further copy (*Chanut*) which he then gave to Clerselier (to which I return in Sect. 1.5). As mentioned above, Van Surck acted as Elisabeth's intermediary with Descartes, so that it is plausible that she could have made a copy of *VanSurck*, even transcribing it personally, as Descartes's letter of 23 November 1646 and Schuyl's words on *Pollot* and *VanSurck* from his 1662 preface suggest.[21] Moreover, I suppose that Elisabeth got her copy after 8 July 1644. Indeed, on that day Descartes wrote to her from Paris touching upon the topic of passions in a way consistent with the treatment he gave of the same topic in his *L'homme*, but without mentioning it,[22] while the first overt mention of his *L'homme* in their correspondence was functional to a treatment of the same topic (in the above-mentioned letter to Elisabeth of 6 October 1645).

Alternatively, let us assume that the second copy of *VanSurck* belonged to Constantijn Huygens (1596–1687). This hypothesis has been advanced by Meschini. Huygens, indeed, after having complained to Descartes on 5 October 1643 about the latter's intention to omit some part of his *Principia philosophiae* (1644),[23] on 23 November 1643 asked him—voicing also the opinion of Pollot—not to omit from his *Principia* the part on man and not to hide it, expressing hope (with Pollot) in obtaining from him a copy of such a "piece," evidently deemed as already written.[24] In fact, as argued by Meschini, Huygens was not directly asking for a copy of *L'homme*; at that time Descartes had written, about man, only such a treatise, and he was never to write a physiological section of his *Principia*, viz. a fifth part on living bodies i.e. animals and plants, and a sixth part on man, notoriously referred to in article 188 of the fourth part of his *Principia*. So that Huygens was nonetheless asking for a text matching the contents of *L'homme*—of which, logically, he was not in possession—and, since he was also expressing interest on it on behalf of Pollot, if Pollot at some point had a copy of *L'homme*, this should have come together with that of Huygens.

Thus, even if Huygens was asking Descartes for the text of his treatment of man also on behalf of Pollot—who logically got his copy of *L'homme* not earlier than November 1643—this does not mean that we can consider Huygens as the owner of a copy drawn from *VanSurck*. After all, (1) Descartes might just have provided Pollot with a copy independently of Huygens's request. Moreover, (2) such a request might have been fulfilled by Descartes by providing Pollot with a copy which Huygens could have access to, even if this contradicts Descartes's recommendation (as declared in his already mentioned letter of 23 November 1646) to the owners of the copies not

---

[21] See *supra*, nn. 5 and 7.
[22] Descartes to Elisabeth, 8 July 1644, in AT V, 64–66.
[23] Huygens to Descartes, 5 October 1643, in AT IV, 756. See Meschini 2011, 180–187.
[24] Huygens to Descartes, 23 November 1643, in AT IV, 766–767.

to show them to anybody,[25] and some evidence, which I discuss in Sect. 1.3, that Huygens was not completely aware of the contents of *L'homme*. As a matter of fact, Elisabeth is in any case a better candidate as the owner of the second copy of *VanSurck*, as there is direct evidence that (1) she owned a copy and that (2) she had been authorized by Descartes to read his *L'homme*.

1.3 Regius

Regius: the existence of this copy is hypothetical, insofar as Henricus Regius (1598–1679) could have had access to Descartes's treatise by reading someone else's copy: though, it is highly probable, being in this case, the unauthorized copy of Descartes's *L'homme* par excellence. His plagiarism of *L'homme* was denounced by Descartes himself in a number of letters commenting upon Regius's *Fundamenta physices* (1646),[26] whose chapters 10 and 12 bear witness to his appropriation of Descartes's theory of the movement of muscles, which has been discussed in a number of studies.[27] After having read Regius's book, indeed, on 5 October 1646 Descartes wrote to Andreas Colvius (1594–1671)—in the attempt to distance his own treatment of the matter from Regius's appropriation and misuse of it—to be sure that Regius had relied on an imperfect and figureless copy of his *L'homme*, so that he could not understand Descartes's theory of the movement of muscles.[28] With the same aim, on 23 November Descartes wrote the aforementioned letter to Mersenne, providing an overview of the dissemination of the treatise authorized by himself: from the messy autograph *Descartes*, then *VanSurck*, on to its unchecked copies *Pollot* and *Elisabeth*, whose owners were forbidden to show them to anybody. Moreover, Descartes declares that he has been informed by someone (probably Huygens, whom he had just met and who was well informed about Regius's work, as I discuss below) that Regius "had, in spite of me, a copy of this writing" only when his *Fundamenta physices* was almost printed, though in a way absolutely obscure to him.[29] The latter claim was probably aimed at not letting anybody think that he could suspect the owners of the authorized copies (*VanSurck*, *Pollot*, *Elisabeth*) as being responsible

---

[25] See *supra*, n. 5.

[26] Descartes to Mersenne, 5 October 1646, in AT IV, 510–511; Descartes to Andreas Colvius, 5 October 1646, in AT IV, 517–518; Descartes to Mersenne, 23 November 1646, in AT IV, 566–567; Descartes to Elisabeth, December 1646, in AT IV, 590; Descartes to Elisabeth, March 1647, in AT IV, 626. See also AT IV, 691, and Descartes's *Lettre-Préface* to Elisabeth opening the 1647 edition of his *Principia*: AT IX/2, 19–20. According to Descartes, Regius did not correctly understand his explanation and prepared a figure contradicting the rules of mechanics, as he relied on an imperfect and figureless copy.

[27] See, for instance, Schmaltz 2016; Strazzoni 2023a (discussing also Regius's apparent appropriations concerning Descartes's theory of hunger and the passions).

[28] Descartes to Colvius, 5 October 1646, in AT IV, 517–518.

[29] Descartes to Mersenne, 23 November 1646, in AT IV, 566–567.

for Regius's appropriation of the treatise, as his recommendations were aimed especially against Regius's possible use of it.[30] As a matter of fact, however, Descartes should have had at least some suspects, though this does not mean that Regius got his copy—which we can safely suppose to be a copy different from the aforementioned ones—directly from Van Surck, Pollot or Elisabeth. My hypothesis is that he got it from Elisabeth, though indirectly: namely through Samson Johnson (1603–1661), whether or not with her permission or with the mediation of further copies.

Johnson, an Englishman, was the Anglican court chaplain of Elizabeth Stuart (i.e. Elisabeth's mother, 1596–1662) until 1644, when he was dismissed from his post as being suspected of Socinianism, though he was in contact with Elisabeth at least until May 1645, before becoming, in 1646, army chaplain at Breda.[31] Together with Elisabeth, Regius, Heereboord, and Henricus Bornius (c. 1617–1675), Johnson was a foremost member of the network of Samuel Sorbière (1617–1670), a main expounder of the ideas of Pierre Gassendi (1592–1655) in the Netherlands. As reconstructed by Vlad Alexandrescu, indeed, it was Johnson himself who introduced Sorbière to Elisabeth in June 1643.[32] In turn, Elisabeth had started her correspondence with Descartes in May 1643 in order to get clarification on an issue raised by the *Physiologia sive Cognitio sanitatis* (1641–1643) of Regius, namely that of the determination of material animal spirits by an immaterial soul, for whose solution Regius himself directed Elisabeth to Descartes.[33] The issue—in the more general terms of the communication between different kinds of substances—was also central to Gassendi's criticisms of Descartes, moved first in his *Obiectiones quintae* to Descartes's *Meditationes de prima philosophia* (1641).[34] In fact, Regius met Sorbière more than once at Utrecht in 1642, giving him a copy of some disputations of his *Physiologia*: a text appreciated by Gassendi, to whom Regius manifested his esteem in 1642.[35] Gassendi and Regius, indeed, came to have in common a number of philosophical positions, manifest especially after summer 1645 (when Regius presented to Descartes a draft version of his *Fundamenta physices*, rejecting Descartes's metaphysics).[36] Johnson, in turn, allegedly abandoned Cartesianism after having read Gassendi's *Disquisitio metaphysica seu Dubitationes et Instantiae* (1644), even if in 1645 he embraced Cartesianism again (it being in any case unclear whether in physics or metaphysics).[37] Both Sorbière and Johnson, moreover, as well as Regius shared Socinian sympathies,

---

[30] See *supra*, n. 5.

[31] Elisabeth to Descartes, 24 May 1645, in AT IV, 207–211. On him, see Hobbes 1994, volume 1, 129–130; Descartes and Regius 2002, 181, n. 4.

[32] Alexandrescu 2012.

[33] Regius 1641–1643, 45; Elisabeth to Descartes, 6 May 1643, in AT III 660–661. For a discussion, see Bos 2017.

[34] AT VII, 339–342; Gassendi 1644, 307–314.

[35] Sorbière to Gassendi, 8 June 1642, in Gassendi 1658, volume 6, 447; Sorbière to Mersenne, 25 August 1642, in Mersenne 1933–1988, volume 11, 240–243; Gassendi to Bornius, 9 August 1646, in Gassendi 1658, volume 6, 253; Sorbière 1691, 210–212.

[36] Fisher 2005.

[37] Sorbière to Gassendi, 10 May 1644, in Gassendi 1658, volume 6, 469–470; Regius to Descartes, 13/23 June 1645, in AT IV, 235.

as testified to by the so-called 'Naarden affair' of 1631, when Regius was accused of Socinianism and Arminianism by the Amsterdam Classis,[38] and by his former disciple Johannes de Raey (c. 1620–1702), to whom I will return below and who attributed to Regius also Epicurean positions—which one may read as a label for Gassendism.[39]

Given these connections, it is not surprising that we can get some insights into Regius's access to Descartes's *L'homme* in 1646 from Sorbière's correspondence network: Sorbière himself wrote about Regius's *Fundamenta physices* to Thomas Hobbes (1588–1679) on 21 May 1646, announcing that it was in course of printing by Louis Elzevier (1604–1670), apparently after Johnson had already promised Hobbes a copy—according to a letter of Hobbes to Sorbière of 1 June, from which it appears that Regius's book was to appear soon. Bornius, in turn, promised on 28 May one or two copies to Gassendi, reiterating his promise on 9 July, when he remarked that its printing was near completion, and that it would have already been finished if Regius had not delayed it.[40] So we can suppose that Regius could have got a copy of *L'homme* between 28 May and 9 July 1646, or slightly before, as it was between these dates that the printing of his *Fundamenta physices* had been delayed. In particular, as revealed by a letter of Sorbière to Hobbes of late September 1646, dealing with the printing of Hobbes's *De cive* (1647), the printing of Hobbes's text had been postponed because Regius sent Elzevier some additions when the latter was already at work on the book, and this caused a "tiresome" delay in the printing.[41] Such additions certainly concerned at least the theory of the movement of muscles, if not hunger and the passions.[42]

As to Johnson, he had an important role in the preparation of Regius's book. There are three main pieces of evidence for this, although indirect. First, once the news started to spread that a book based on Descartes's principles was going to be published, Mersenne asked Descartes (in summer 1646) whether Johnson was its author.[43] Second, Huygens directed to Johnson himself a reprimand on the book, somehow kicking the dog while meaning the master (Regius). After having fulfilled

---

[38] De Vrijer 1917, appendix 2.

[39] Borch 1983, volume 1, 43.

[40] Sorbière to Hobbes, 11/21 May 1646, in Hobbes 1994, volume 1, 128–129; Bornius to Gassendi, 28 May 1646, in Gassendi 1658, volume 6, 498–499; Hobbes to Sorbière, 1 June 1646, in Hobbes 1994, volume 1, 131–132; Bornius to Gassendi, 9 July 1646, in Gassendi 1658, volume 6, 499.

[41] Sorbière to Hobbes, late September 1646, in Hobbes 1994, volume 1, 136. Regius's book finished being printed between 21 August and 1 September, when, respectively, Regius dated his dedicatory letter to the Stadtholder and sent Huygens three complimentary copies of his book – one of which was for the Stadtholder: Regius 1646, *Frederico Henrico* (unnumbered), 10/21 August 1646; Regius to Huygens, 1 September 1646, Amsterdam, Universiteitsbibliotheek, ms. K 137, partially transcribed in Huygens 1911–1917, volume 4, 346.

[42] See Strazzoni 2023a.

[43] Descartes to Mersenne, 7 September 1646, in AT IV, 497. Descartes answers to a letter of Mersenne, now lost, written before he could read a letter of Huygens of 21 August, where details on Regius's *Fundamenta physices* are provided: Huygens to Mersenne, 21 August 1646: Mersenne 1933–1988, volume 14, 413.

1.3 Regius

Regius's request of presenting the book to the Stadtholder Frederik Hendrik (1584–1647),[44] Huygens nonetheless harshly complained to Johnson, to whom he wrote on 27 September to have the news of the presentation forwarded to Regius himself, about the lack of credit due to Descartes by Regius, "of which, without Descartes (...), not even a syllable could have come into his [Regius's] mind."[45] Notably, the letter offers evidence that Huygens was aware that Regius had access to Descartes's *L'homme*, and at the same time that this was not direct access by Huygens himself, viz. he was not the owner of the copy twin to *Pollot*, as mentioned in Sect. 1.2. Indeed, with the physiological chapters (8–10) of his *Fundamenta physices*, Regius ideally completed the above-mentioned plan set by Descartes for his *Principia*. In these chapters, even if plagiarizing Descartes at some point, Regius presented (as he already had in his *Physiologia*) a whole theory of physiology which he had developed on his own. Huygens, in turn, seems to ignore this, as he does not concede any originality to Regius. Third, Johnson's role in the preparation of Regius's *Fundamenta physices* is confirmed by Elisabeth's letter to Descartes of 11 April 1647. In it, Elisabeth reveals that Regius himself told her that he was assisted by Johnson in writing his book (and this communication took place no later than July 1646, when she moved to Germany).[46]

Finally, Elisabeth's role in Regius's access to a copy of *L'homme* can be corroborated by considering that some ideas which Regius appropriated from Descartes's treatise concerned the passions,[47] a topic which, as mentioned in Sect. 1.2, was also central to Elisabeth's correspondence with Descartes,[48] and for the discussion of which Elisabeth could rely on *L'homme*: so that Elisabeth might have helped Regius—although indirectly—to deal with the same topic by using *L'homme*. Moreover, we can consider that in any case at some point she contravened Descartes's recommendations not to disseminate the treatise further. As seen in Sect. 1.2, indeed, she provided La Voyette with a copy, and this probably happened when they were both in the Netherlands, namely up to July 1646, before the publication of Regius's *Fundamenta physices*, when Descartes's recommendations on the circulation of his *L'homme* were still in force. Certainly, La Voyette met Elisabeth in The Hague, where they both resided and belonged to the Cartesian circle,[49] before she left for Germany

---

[44] See *supra*, n. 41.

[45] Huygens to Johnson, 27 September 1646, The Hague, Koninklijke Bibliotheek, ms. KA 45, 141v; partially transcribed in Huygens 1911–1917, volume 4, 354. No reaction to Huygens's criticism is, in any case, revealed by Regius's subsequent letter to him of 28 November/8 December 1646 (Uppsala, Universitetsbiblioteket, Waller Ms benl-00586), reporting that he had been kept informed by Johnson and Bornius about the success of the presentation of the book and its dedication to the Stadtholder.

[46] Elisabeth to Descartes, 11 April 1647, in AT IV, 630.

[47] As discussed in Strazzoni 2023a.

[48] Muller 2023.

[49] On him, see Huygens 1892–1899, volume 4, 209–214; Huygens to Christopher Delphicus von Dohna, 24 July 1653, in Huygens 1911–1917, volume 5, 180–181; Ulfeldt 1949, 79; Tessin 1965, 59 and 263; Jacobson and Hildebrand 1945; Descartes 2003, xi–xv.

in July 1646,[50] after which she never came back to the Netherlands. We cannot exclude that she provided him with a copy in or from Germany after 1646, in person or with the intermediation of somebody else, like his friend Christopher Delphicus von Dohna (1628–1668), who was himself a Cartesian and a member of Sorbière's network.[51] However, given the fact that La Voyette stayed in the Netherlands at least until 1650, while from mid-1651 he was in Sweden (at least until 1653, serving in the same year as captain in the Life Guards of Christina of Sweden, 1626–1689)—while Elisabeth was in Germany—it is probable that he got his copy from Elisabeth no later than 1646. In turn, Chevreau had access to the treatise in early 1653, as he moved to Stockholm from London around March of that year, and certainly met La Voyette there.[52]

1.4 Other Dutch Copies

1.4.1 Heereboord

Notably, Regius—besides Elisabeth—was not the only member of Sorbière's circle owning a copy (or at least having access to the text): indeed, Heereboord, who was a notable member of the circle (as well as being mentor to Bornius and an admirer of Gassendi)[53] had one (*Heereboord*). In fact, Heereboord had been interested in obtaining a copy as early as 1642, when he got acquainted with the existence of Descartes's treatise through Adolph Vorstius (1597–1663), and asked Colvius to intercede with Descartes for a copy.[54] As mentioned in Sect. 1.2, the existence of his copy is testified to by Schuyl, who in his 1664 *Epistola ad amicum*, claims that Descartes's use of the term 'tuyau' (pipe or tube) is testified to by a copy that Heereboord had lent him – as well as by the two manuscripts mentioned in the 1662 preface, viz. *Schuyl* (or *Pollot*) and *VanSurck*. Indeed, Schuyl's translation of this term in his 1662 edition had been criticized by the undisclosed recipient of Schuyl's *Epistola*, who can be identified in De Raey (as I discuss in Sect. 1.4.2).[55]

---

[50] Descartes to Elisabeth, July 1646, in AT IV, 448–449.

[51] On him, see Jacobson and Hildebrand 1945; see also Baillet 1691, volume 2, 297; Sorbière to Elisabeth, 3 June 1652, in Sorbière 1660, 69–77; Sorbière to Dohna, 3 June 1652, in Sorbière 1660, 77–80; Lanfrey 1879, 67.

[52] Boissière 1909, 11; Jacobson and Hildebrand 1945.

[53] Del Prete 2020.

[54] Heereboord to Colvius, 8 April 1642, in AT VIII/2, 196. Van Otegem has taken this letter as evidence that one of the two copies of *VanSurck* was *Heereboord*; in turn, Meschini has argued that the letter shows, at most, that in 1642 Heereboord was not in possession of a copy: Van Otegem 2002, volume 2, 487; Meschini 2011, 172–173.

[55] "(…) *tuyaux*, quam Des-Cartes iis in locis usurpavit; uti mihi quatuor testantur exemplaria: nimirum duo accuratissima, quorum mentio facta est in Praefatione, et aliud, quod Clariss. D. Heerebortius mihi commodaverat. Et denique exemplar, ex authoris autographo accurate descriptum, quod

1.4 Other Dutch Copies

Heereboord probably lent his copy to Schuyl in 1659–1661, before dying on 7 July of the latter year: probably because—as suggested above—Schuyl was attempting to make sense of Descartes's text and was looking for additional copies, after having had access to *Schuyl* (before or in early 1659), and before reverting to *VanSurck*. Notably, as Schuyl, in his 1664 *Epistola*, claims to have asked for a check on 'tuyau' in a fourth copy (to which I will return in Sect. 1.4.2) of the treatise, it might be that at that point he had no more access to *Heereboord* and *VanSurck* (while *Schuyl* certainly belonged to him). Still, as (1) he uses the present tense in claiming that four manuscripts testify to Descartes's use of the term 'tuyau';[56] (2) as *VanSurck* could have been a gift to him, to allow him to finish his work (as seen in Sect. 1.2), and (3) as *VanSurck* (not to mention *Heereboord*) had probably been consulted to a lesser extent than *Schuyl* for the editing of the text, and so was physically needed for a check in late 1663–early 1664, we can suppose that both copies (*Heereboord* and *VanSurck*) were still at Schuyl's disposal at that time, without *Heereboord* having been returned before its owner's death. We can in any case exclude that Heereboord lent it to Schuyl in order to conduct such a check: indeed, as I discuss in Sect. 1.4.2, De Raey commented upon a published version of Schuyl's translation i.e. after summer 1662, when it came to light, and Schuyl got acquainted with his criticisms in late 1663–early 1664.[57] This does not exclude that Schuyl received Heereboord's copy when it was too late to be used for his translation: this is unlikely, though, as his translation was published circa one year after Heereboord's death. As mentioned in Sect. 1.2, in fact, *Heereboord* was probably of a quality inferior to the copies of Pollot and Van Surck, who both motivated Schuyl to work on his edition (a third one being Clerselier, who apparently encouraged Schuyl in 1660–1661, after the latter got *Schuyl* and *VanSurck*).[58] This suggests that *Heereboord* was not one of the direct copies of *VanSurck*, i.e. that it was part of a further copying of the treatise—probably originating from *Regius* (as I argue further in Sect. 1.4.2)—and that Heereboord had no role in Schuyl's editorial project (and indeed his psycho-physical condition had deteriorated already in 1652).[59]

1.4.2 De Raey

As in the case of Regius's copy, the existence of a copy belonging to De Raey (*DeRaey*) as a separate one is also hypothetical, though there is abundant evidence that he had access to the text over the years. Such a copy most probably originated

---

Ampliss. D. Clerselier Celeberrimo D. Gutshovio, in Academia Lovaniensi Anatomices et Matheseos Professori Regio, Parisiis transmisit," Descartes 1664a, *Intepretis epistola ad amicum*, 1–2 (unnumbered); italics by Schuyl. As to the fourth copy, see Sect. 1.4.2.

[56] See *supra*, n. 55.
[57] See *supra*, n. 15, and *infra*, n. 80.
[58] See *supra*, n. 7, and *infra*, n. 104.
[59] Verbeek 2003; Strazzoni 2014, 81–82.

from a member of the circle of Sorbière. Besides being a personal acquaintance of Descartes and a witness at the opening of his trunk at Leiden in March 1650, De Raey had indeed been a student both of Regius at Utrecht in 1641–1643 (where he acted as *respondens* in his *Physiologia*) and of Heereboord at Leiden in 1643–1647 (where he was also a student of Vorstius), with whom he kept ongoing contact well after his days as a student, and becoming, in 1647, a colleague of Heereboord and Vorstius at Leiden.[60] If we suppose that, as Descartes put it, Regius "could not stop himself to tell his disciples about it [viz. his *L'homme*] if he had seen it,"[61] we can hypothesize that De Raey got his copy from his teacher Regius, though it is more probable that he got it between 1652 and 1656, well after his student days, nonetheless probably deriving from *Regius* and/or *Heereboord*, if not intermediate between the two, with De Raey acting between Utrecht and Leiden.[62] In any case, we can exclude that De Raey was in possession of an autograph (either from Descartes's Leiden trunk or from any other source), since, as I show in Sect. 1.5, Clerselier was in possession of an autograph, and De Raey, who notoriously was a fierce criticizer of the French 'appropriation' of Descartes,[63] certainly did not send it to him.

We know that De Raey owned or had access to a copy from the contents of his academic lectures (*dictata*), bearing evidence to his reliance on some contents of Descartes's treatise. Namely: (1) a commentary on the *Principia* tracing to 1658 and incorporating Descartes's account of sense perception;[64] (2) a commentary on the *Epitome Institutionum medicarum* (1631) of Daniel Sennert (1572–1637),

---

[60] Strazzoni 2022. De Raey was well known in Sorbière's circle: Sorbière to Bornius, 10 September 1652, in Sorbière 1673, volume 1, 212r.

[61] Descartes to Elisabeth, March 1647, in AT IV, 626.

[62] See *infra,* nn. 65 and 169.

[63] Baillet 1691, volume 1, xxix–xxxii; Ritter 1705, 4 (unnumbered).

[64] In his *Annotata ad Principia philosophica* (1658), De Raey, commenting upon *Principia* IV.191, explains sense perception as such as the effect of the movement of the filaments of nerves, which, being pulled but not broken, open the pores of the substance of the brain, causing the animal spirits to flow towards the part of the body where we feel a certain sensation: "(…) ex qua tensione et tractione pori in cerebro valde aperiuntur, et vehementer scaturiunt animales spiritus, atque eo tendunt, quod proinde vehementer sensationem excitat," De Raey 1658, 564. While the idea that sense perception happens by the pulling of filaments which in turn move certain parts of the brain is briefly expressed in the *Dioptrique* (1637; AT VI, 111) and *Meditationes* (AT VII, 87), the idea that it happens through the dilatation of pores and the increased flowing of spirits in them is expressed only in Descartes's *L'homme* (in all its versions: Descartes 1662, 35–35; Descartes 1664b, 29–30; ATH 1444, 33–34). On De Raey's *dictata* and their dating, see Strazzoni 2023b.

1.4 Other Dutch Copies   17

dating to 1656 or 1661 (also extant in an undated copy) and presenting *L'homme*-inspired accounts of the generation of animal spirits, the pineal gland,[65] digestion, and hunger.[66]

Moreover, evidence on his access to the text can be gathered from two other interrelated sources. First, from a diary entry of Ole Borch (1626–1690) of 22 October 1662, reporting that on the occasion of the funeral ceremonies of the orientalist Johann Georg Nissel (1623–1662), De Raey told Borch that no page of Schuyl's translation was exempt from error, as Schuyl did not understand French. In particular, De Raey criticized Schuyl's translation of the term 'maille' (in its only occurrence in the treatise, in Clerselier's version)—thereby revealing that he had access to a French text before the appearance of Clerselier's 1664 edition—by 'funiculus' (filament,

---

[65] De Raey provides an account matching that of *L'homme*, and which cannot be found in other places where he discusses such topics, as in his *Discours de la méthode* (1637; part 5), *Les passions de l'âme* (1649; articles 31 and 32), or in his correspondence: AT VI, 54–55; AT XI, 351–353; Descartes to Mersenne, 24 December and 30 July 1640: AT III, 136 and 263). According to De Raey, spirits exhale in the cavities of the brain filtered by the membranes of the arteries and the substance of the glands: namely through (1) the arteries of the *rete mirabile*, and (2) those of the choroid plexus (the only part not mentioned by Descartes), which envelop the (3) pineal gland and attach it to the brain; the pineal gland itself is described as a "spring of spirits": "spiritus, qui tanquam potior pars sanguinis una cum sanguine in corde sunt generati, in cerebro tantum segregari et exhalando produci. Cumque sanguis arteriosus non extravasetur in cerebro, necesse est spiritum exhalare per arteriarum membranas et forte colari per glandularum densam substantiam, quod ordinarium naturae artificium est in secretionibus. Arteriolae autem hae duobus in locis notabiles sunt, circa glandulam pituitariam, ubi rete admirabile constituunt, et in cerebri ventriculis, ubi plexum choroidis faciunt, qui etiam versus glandulam pinealem ascendit, eamque multis subtilissimis arteriolis annectit et omni ex parte cingit, quod a Cartesio ante multos annos primo notatum, (…). [C]onsideremus ex illa et circa illam etiam spirituum scaturiginem esse, ita ut illud etiam sit organum motus et revera particula illas de qua Aristoteles lib. De animal. mot. dicit, quod facta parva eius commotione (haud aliter atque per gubernaculum in navi) magna inde mutatio sequatur in corpore reliquo," De Raey 1656/1661, 16–17 and 33; cf. Descartes 1662, 14–18; Descartes 1664b, 10–13; ATH 1444, 13–18; Aristotle, *De motu animalium*, part 7, 701b 26–28. The metaphor of the oar and the reference to Aristotle are also present in a disputation by De Raey of 6 December 1651 (re-issued in his *Clavis philosophiae naturalis*, 1654: De Raey 1654, 75), though without any trace of his access to *L'homme*, which thus most probably happened after that year (see also *infra*, n. 169). Moreover, an associate of De Raey had access to the treatise no later than 1652 and certainly in 1656, as I discuss in Sect. 1.4.3: at which point De Raey should also have had access to it. See also *infra*, n. 170.

[66] As in the case of De Raey's exemplification of the transformation of chyle into blood ("chylus eo fere modo se habet ad sanguinem, ut mustum ad vinum clarum et bene fermentatum," De Raey 1656/1661, 16; cf. Descartes 1662, 4; Descartes 1664b, 4; ATH 1444, 4), his explanation of hunger and thirstiness ("eadem est causa famis, quam coctionis, humor nimirum subtilis et acris qui in corpore nostro nunquam est otiosus, adeoque in ventriculo a cibis vacuo, vel non satis repleto, non inveniens in quod agat, vim suam exerit in nervosam ventriculi substantiam. Sitis proxima causa est siccitas oris, quae a defectu humorum, vel nimio fervore oritur, vel ab utraque causa," De Raey 1656/1661, 19; cf. Descartes 1662, 65–67; Descartes 1664b, 55–57; ATH 1444, 63–64), and the identification of three causes of concoction ("verae coctionis causae sunt: 1. spontanea alimentorum putrefatio seu digestio, (…) 2. causa coctionis vere principalis est ad cibi in satis subtiles particulas resolutionem brevi tempore peragendam necessaria, subtilis nempe et insigni vi penetrandi ac fermentandi praeditus liquor, qui ex sanguine arterioso in ventriculi capacitatem continuo exhalat. 3. Denique accedit calor actualis," De Raey 1656/1661, 20; cf. Descartes 1662, 3; Descartes 1664b, 2–3; ATH 1444, 2–3).

rope), which for him should have been rather translated by 'intervallum' (opening). The reference is to the constitution of the brain according to Descartes, for whom it is composed of filaments constituting a mesh or *maille*, which is in fact an ambiguous term (to be very generous with Schuyl), as by 'mailles' one can intend the very filaments or ropes composing a net, or the spaces between them,[67] namely the same structure that he also rendered with the term 'intervalle' in Clerselier's edition, and translated with 'intervallum' in Schuyl's version and in ATH 1444.[68] According to De Raey, it was because of this error that Schuyl deemed the filaments of the brain (*filets* or *fibrillae*, in the original and in Schuyl's translation) to be hollow, against Descartes's tenet that such filaments were solid and not hollow, and that between them there are hollow interstices, namely the above-mentioned *mailles*.[69] Notably, De Raey himself (evidently discontent with Schuyl's translation) prepared a Latin translation of Descartes's *L'homme*. According to the sale catalogue (1723) of the private library of Johann Theodor Schalbruch (1655–1723), friend and colleague of De Raey at Amsterdam, De Raey was in fact the author of a "handwritten evidently new translation" of Descartes's *L'homme*, which was added to or bound with a copy of the 1664 Latin edition of Schuyl.[70] Even if we cannot assume that he made his translation on the basis of a manuscript and not of Clerselier's edition, his attitude towards the French appropriation of Descartes make probable that he used his copy (*DeRaey*) of the treatise. A translation that, in any case, was certainly different from the one provided in ATH 1444: as I discuss in Sect. 1.7, indeed, the text of ATH 1444 is vulnerable to De Raey's criticisms even more than Schuyl's translation.

Borch's diary entry allows us to identify in De Raey the 'friend' addressed by Schuyl in the above-mentioned 1664 *Epistola*, which is the second main source identifying him as a possessor of a copy. In his *Epistola*, Schuyl addresses, first, what is the second criticism expounded by De Raey to Borch. Namely, Schuyl replies to the *amicus*'s criticism (which came to Schuyl apparently through an annotated copy of the 1662 edition, as Schuyl refers to deleted and replaced words)[71] of his translation

---

[67] "(…) pour ce qui est des pores du cerveau, ils ne doivent pas être imaginés autrement que comme les intervalles qui se trouvent entre les filets de quelque tissu (…). Concevez sa superficie (…) comme une résille ou lacis assez épais et pressé, dont toutes les mailles sont autant de petits tuyaux par où les esprits animaux peuvent entrer," Descartes 1664b, 62–63; cf. Descartes 1662, 73: "[c]uius singula filamenta nihil aliud sunt, quam totidem tubi." See Sect. 1.7 for a discussion of the equivalent place in ATH 1444.

[68] Cf. Descartes 1662, 72, 74, and 83; Descartes 1664b, 62, 64, and 74; ATH 1444, 71, 72, and 79. As I discuss in Sect. 1.7, in ATH 1444 there is in any case provided a different characterization of the *intervalla*.

[69] "Eodem die exeqvias ii Nisselio, ubi inter colloqvia narravit D. de Raei in Scÿlij versione Cartesianâ de homine ne unum folium esse sine labe, namqve Scÿlium non intellexisse ita præcisè Gallica, e.g. mallie, qvod esset reddendum, intervalla, reddidisse per funiculos, adeoqve statuere fibrillas cerebri esse cavas, contra mentem Cartesij, qvi vult fibrillas cerebri esse solidas, interstitia a. inter fibrillas, ille vocat mallie, i.e. intervalla cava," Borch 1983, volume 2, 217.

[70] "Tractatus de homine, Latinitate donatus a Flor. Schuyl. Lugd. Bat. 1664. Accessit eiusdem versio plane nova ms. auctore Jo. de Raei," *Catalogus* 1723, 121.

[71] This is revealed by a quite trivial remark by the *amicus* viz. De Raey, addressed by Schuyl in the second place: "[d]enique quod, ubi ego *per transennam* posui, pag. 13, ille hoc deleverit, atque

1.4 Other Dutch Copies

of 'tuyaux' with 'tubuli' (viz. small tubes) in two occurrences of this term in Schuyl's 1662 edition, suggesting instead that these have to be meant not as hollow structures and translated by 'filamenta' (viz. filaments).[72]

After having received or being acquainted with De Raey's remarks (which probably happened only in late 1663–early 1664),[73] Schuyl made a request—as anticipated in Sect. 1.4.1—to make a check on such a word ('tuyau'), to Gerard van Gutschoven (1615–1668), who was one of the two ascertained illustrators of Clerselier's edition (the other being Louis de La Forge, 1632–1666). Van Gutschoven checked such recurrences in the copy (*VanGutschoven*) which Clerselier had provided to him, around summer 1660, to prepare the figures (as he did also with La Forge in the same period: *LaForge*).[74] Van Gutschoven's copy (like *VanSurck*, *Schuyl* (or *Pollot*), and *Heereboord*), contained 'tuyaux' (as reported by Van Gutschoven to Schuyl in a letter quoted by the latter),[75] so that Schuyl rebukes the *amicus*'s corrections as evidently unwarranted. What is at stake is Descartes's description of the structure of the nerves. According to all the extant versions, nerves are in fact unmistakably composed of (1) a membrane originating from a membrane enveloping the brain and forming a big tube; this big tube contains (2) other smaller tubes, originating from another membrane enveloping the brain (all such tubes being filled by the animal spirits flowing from the inner cavity of the brain towards the other parts of the body, for instance in order to fill the muscles and move the body). Such smaller tubes, eventually, contain (3) some filaments originating from the inner substance

---

reposuerit *obiter*, aut alibi saepe quid simile mutandum duxerit; id illius delectui tribuendum," Descartes 1664a, *Intepretis epistola ad amicum*, 3 (unnumbered); italics by Schuyl; cf. Descartes 1664b, 10: "en passant." See Sect. 1.7 for a discussion of the equivalent place in ATH 1444.

[72] "[E]rrat, et duplici, uti opinor, nomine, quisquis ille sit, qui pag. 19, lin. 7 libelli De homine, negat legendum esse: *Ecce itaque, exempli gratia, hunc nervum A, cuius exterior membrana, sive tunica, tubo satis amplo similis est, suoque ambitu plures alios tubulos b, c, k, l, et c. complectitur.* Sed *tubulorum* loco *filamenta* esse reponenda censet: atque etiam pag. 21, lin. 10, delenda putat haec verba, *isti tubuli*, et reponenda, *ista filamenta*, quae quidem non ut *tubulos*, sed ut *solida corpora* consideranda putat," Descartes 1664a, *Intepretis epistola ad amicum*, 1 (unnumbered); italics by Schuyl.

[73] See *infra*, n. 80.

[74] Descartes 1664b, *Préface*, 5–13 (unnumbered). On the lengthy delivery of a copy to Van Gutschoven (with which Clerselier had reached an agreement in late 1659, with *VanGutschoven* thus prepared in 1659–1660) and on the preparation of the figures for Clerselier's edition, see Strazzoni 2023a. A copy, that of Van Gutschoven, which was most probably also the source of another one (*Philippi*), or which was at least shown to his colleague at Leuven Guillaume Philippi, who in his *Medulla logicae* (1661) provided an explanation of hunger textually relying on Descartes's *L'homme*: the textual agreements are presented in Monchamp 1886, 329–331.

[75] "Quare et eius verba ex illius ad me cursim data epistola hic lubet adscribere. I. *Quod pag. 19 lin. 7 hisce verbis expressisti*, suoque ambitu plures alios tubulos b, c, k, l, et c. complectitur, *ita Gallice habet Cartesius*, qui contient plusieurs autre petits tuyaux b, c, k, l, et c. *Ex quibus verbis id bene translatum iudico*, et nullo modo filamenta *loco* tubulorum *esse ponenda*. II. p. 21 *non* ista filamenta, *sed* isti tubuli, *uti habes, legendum est. Haec enim sunt in tractatu Gallico Cartesii verba*: ou ces petits tuyaux, qui et c. Huc usque doctissimus D. Gutishovius," Descartes 1664a, *Intepretis epistola ad amicum*, 2 (unnumbered); italics by Schuyl.

of the brain, constituting a sort of marrow.[76] No doubt, in Schuyl's reconstruction, Descartes's reference to "other tubes" clearly differentiated between a containing tube and contained smaller tubes—while De Raey claimed that such smaller tubes were in fact not hollow filaments.[77]

How to make sense, therefore, of De Raey's apparently unwarranted criticism? First of all, let us assume—as Schuyl does in his *Epistola*—that De Raey might have been deceived by the fact that the French word 'tuyau' is akin to the Dutch word 'touw', meaning rope or filament, or that he was relying on a badly transcribed copy.[78] A possible objection to the latter hypothesis is that De Raey could have had access to a copy provided by Clerselier to one of his associates around 1657, namely Aernout Huyberts (1633–1716), whose efforts I discuss in Sect. 1.4.3. However, considering De Raey's attitude towards the French, it is still possible that he just did not deem such a copy reliable. As a third hypothesis, let us assume that De Raey nonetheless had at his disposal a text identical to the other extant ones with regard to the discussion of the constitution of the nerves: to be generous to De Raey, these texts are not entirely clear on how the nerves outside the brain connect with the internal structure of the brain, which does not contain complete nerves, as pointed out by De Raey himself in his *dictata*, where he just remarks on the role of spirits in connecting the nerves and the pineal gland.[79] This supposition can be corroborated by Schuyl's *Epistola*, where he concedes one error noticed by De Raey, but at the same time offers some interpretations of the text—with regard to the structure of the nerves and the brain—which are at best objectionable, on the ground of his very version of the text. Schuyl accepted (at the end of his *Epistola*) the first, above-mentioned criticism moved by De Raey, as he reports to have told the publisher, albeit too late to be

---

[76] Cf. Descartes 1662, 19–21; Descartes 1664b, 15; ATH 1444, 19–20.

[77] "Certe relatio, qua pag. 19 *tubus* ille *amplus* refertur ad *plures alios tubulos*, praeter verbi proprietatem mentem authoris abunde declarat," Descartes 1664a, *Intepretis epistola ad amicum*, 3 (unnumbered); italics by Schuyl. Also the other correction, suggested by De Raey on page 21, line 10 ("isti tubuli" to be replaced with "ista filamenta") clearly contradicts Descartes's argumentation: Descartes 1664a, *Intepretis epistola ad amicum*, 2 (unnumbered).

[78] Descartes 1664a, *Intepretis epistola ad amicum*, 3 (unnumbered). At that point De Raey might have used Descartes's *Dioptrique* and *Passions* for reading *L'homme*, as in both these published texts Descartes proposes a simplified structure of nerves, according to which they are composed of (1) the same membranes enveloping the brain, constituting their skin, (2) a marrow of the same substance as that of the brain, constituting the filaments: AT VI, 109–110; AT XI, 337. A marginal annotation (by the same hand of the main text) to one of his *dictata* on Sennert endorses such a simplified structure: "[n]ota nervos ex tenuissimis constare membranulis, in sua cavitate medullam continentes, quae undique spiritibus instar venti, vel aeris subtilissimi circumfusae sunt, haecque medulla interior inservit sensibus, motui vero spiritus," De Raey 1656/1661, 79.

[79] "At non terminantur nervi in toto cerebro, sed circa principium cerebri, cerebelli ac spinalis medullae. Non tamen in iis ipsis nervorum finibus immediate fiunt sensus interni, quia in omnium sensuum perceptionibus anima utitur ministerio spirituum, quatenus illi, vel ab impressionibus ab obiecto praesente ortis, vel a praeteritarum impressionum vestigiis afficitur ac patitur, easque suas affectiones et passiones in uno quodam et communi omnium functionum animalium organo repraesentat, quod commune organum undiquaque spiritibus cinctum atque in medio cerebri positum esse oportet," De Raey 1656/1661, 33.

included in the 1664 edition of his translation,[80] to replace the expression 'singula filamenta', used for the translation of "dont tous les mailles" (i.e. the interstices or pores between the filaments constituting the inner substance of the brain), with the expression 'singuli pori'. A correction which was approved by Van Gutschoven, who suggested similar expressions for it.[81] This notwithstanding, Schuyl, first, did not understand why De Raey did not correct some of the occurrences of the term 'tubulus' on pages of his translation in which such a term was used by Schuyl to translate the term 'tuyau', which for Schuyl was used by Descartes to indicate, at this place, the same structure named 'tuyau' in the passages elsewhere corrected by De Raey.[82] Such are pages 79, 80, 89, and 97 of Schuyl's 1662 edition: even if on page 79 there are no occurrences of such a word (this probably being a typo in the *Epistola*), on the other pages—viz. in the French from which Schuyl translated, and in any case in Clerselier's corresponding edition—Descartes uses 'tuyaux' to indicate the very *mailles*,[83] previously clearly compared by him to tubes.[84]

Second, Schuyl claims, in his *Epistola*, that in some cases Descartes used the term 'filet' while, if one looks at the function and source of the structure called in this way, he should in fact have used 'tuyau', viz. Schuyl claims that such *filets* have to be considered as hollow.[85] He mentions an occurrence where we do find a slight variant between the version of Schuyl (containing "filamenta illa") and those of Clerselier and ATH 1444 (which contain "les" and "illa" respectively), though it is clear from all the versions of the paragraph of that occurrence that, after having differentiated the three main structures of the nerve (*grand tuyau/tubus amplus, petits tuyaux/tubuli, and filets/filamenta*), Descartes maintains that the filaments of the nerves have two

---

[80] Thus, he got acquainted with De Raey's criticisms only in late 1663–early 1664, as the *Epistola* is dated 6 March 1664 (Descartes 1664a, *Intepretis epistola ad amicum*, 4, unnumbered), and is not even bound in all the copies of Schuyl's 1664 *De homine*. Also, Schuyl reports to have "hastily" received Van Gutschoven's letter, so that the matter had to be settled in a short time: see *supra*, n. 75.

[81] "Scripseram, sed sero, ad typographum, ut, ubi Des-Cartes habet, *Dont tous les mailles*, p. 73, lin. 4, non poneret, per *singula filamenta*, quamvis id recte explicari posset, sed, per *singulos poros*, quod tamen D. Gutshovius accuratius explicandum putat per *singula cancellorum*, sive *retis foramina*," Descartes 1664a, *Intepretis epistola ad amicum*, 3–4 (unnumbered); italics by Schuyl.

[82] "Et vero, quod non minus ad rem facere videtur, Des Cartes indefinitis locis eadem illa organa, quae maiori nervorum membrana continentur, quasque censor iste locis paulo ante citatis solida filamenta vocanda putat, *tuyaux*, id est, *tubulos* appellat. Uti, praeter alia innumera loca, pag. 79, 80, 89, 97, etc. Quandoquidem igitur ipse censor iis locis *tuyaux*, per *tubulos* recte verti iudicat, quibus eadem illa res designatur, quam pag. 19 et 21, non nisi per *solida filamenta* vult denotatam," Descartes 1664a, *Intepretis epistola ad amicum*, 3–4 (unnumbered); italics by Schuyl. The other corrections by De Raey are those indicated *supra*, nn. 72 and 77.

[83] Cf. Descartes 1662, 80 (7 occurrences), 89 (3 occurrences), and 97 (1 occurrence); Descartes 1664b, 71–72, 78, and 85; ATH 1444, 76, 83, and 90.

[84] See *supra*, n. 67.

[85] "Non diffiteor Cartesium eadem haec organa quandoque *filets*, id est, *filamenta* vocare. Uti liquet ex pag. 21, lin. 4, et alibi. Sed ubi Des-Cartes habet *filets*, ego *filamenta*, ubi vero *tuyaux*, *tubulos* scribendum esse reputavi. Atque exinde existimavi haec esse tubulos instar filamentorum, aut filamenta cava. Uti etiam eorum origo atque officium commonstrare videntur," Descartes 1664a, *Intepretis epistola ad amicum*, 3 (unnumbered); italics by Schuyl.

extremities: one on the internal surface of the brain, facing its concavities, and the other in the external parts of the body where the small tubes containing them ("les"/ "illa") i.e. those filaments ("filamenta illa") end.[86] An interpretation, by Schuyl, which is no less surprising, prima facie, than that of De Raey of the same paragraph.

In conclusion, De Raey was not wrong in claiming that Schuyl mistook filaments and tubes. This, however, might have derived from the fact that Descartes had omitted to clarify how nerves and *mailles* connect—i.e. from the fact that his text(s) had to be interpreted in the light of anatomical observations exploring, amongst others, this missing link and more generally the constitution of the brain. This was noticed by Schuyl himself, who claimed in his 1662 preface that it would be possible to conceive nerves as being closer to the pineal gland than conceived by Descartes, and to intend the nerves thus internal to the brain as connected, through filaments, to the external ones, as in fact he himself does in one of the figures provided to illustrate the constitution of nerves (Fig. A.1 in the Appendix): this still being consistent with Descartes's physiology.[87] Accordingly, De Raey's apparently inconsistent criticism might have derived both from his relying on a defective copy, as well as from the inherent difficulties in interpreting Descartes's text, regardless of the copy he used. I corroborate this interpretation in what follows.

1.4.3 Huyberts

We know that the above-mentioned Huyberts was in possession of a copy (*Huyberts1*) from Clerselier's preface to his 1664 edition of Descartes's *L'homme*. In it, Clerselier declares that in 1657, just after the printing of the first volume of his 1657–1667 edition of Descartes's correspondence (finished being printed on 30 January), he was informed by Louis and Daniel Elzevier (1626–1680, who in 1655 had joined his cousin Louis at the Amsterdam establishment)[88] that a certain Mr. Huyberts had already worked on and completed all the figures of Descartes's *L'homme*. Evidently, these were for an edition of this treatise to be published by the Elzeviers, who asked Clerselier to send Huyberts a copy of the autograph or "original" they had heard was in his possession (which I discuss in Sect. 1.5), in order to check whether Huyberts had done good work (on which they had expressed reservations also in 1656, as I discuss in Sect. 1.6) and to resolve some difficulty related to it, given the fact that the

---

[86] Descartes 1662, 19–21; Descartes 1664b, 15; ATH 1444, 19–20.

[87] "[N]on admodum repugnarem, si quis forte nervos propius ad glandulam produceret: nervosque internos, filamentis quibusdam annecteret externis (…), forte uti nos pag. 20. (…) Si modo omnes corporei motus humani corporis, atque adeo cuiuslibet animalis, ex eius hypothesis commode deduci possent, et in thesi cum authore conveniret" Descartes 1662, *Ad lectorem*, 31 (unnumbered). Though also in the representations of the brain by Van Gutschoven and La Forge one can notice tubular structures entering the brain: see, for instance, Descartes 1664b, 62–65, 67–71, and 74. In a variant present in Schuyl's version (signalled in the present edition), moreover, the intervals between the filaments of the brain are designated as pores of the nerves: Descartes 1662, 68.

[88] Davies 1954, chapter 6.

manuscript used by Huyberts was "not very correct in some places." So Clerselier, after having made Huyberts promise to share with him his figures, sent him a copy (*Huyberts2*) of his autograph, which Huyberts confirmed receiving (probably in 1657). However, afterwards Clerselier heard nothing more from him, explaining this on account of Huyberts's (supposed) condition of serious illness.[89]

Up to now, this was all that we knew about Huyberts as a possessor of a copy of Descartes's *L'homme*. However, Huyberts was a well-known figure in Dutch intellectual circles in the second half of the seventeenth century, and his name surfaces in a number of other sources. Born in Amsterdam in 1633, and enrolled as a student of philosophy at Leiden in May 1649, graduating in medicine on 19 December 1652 under Vorstius, with a *Disputatio de affectibus ventriculi circa concoctionem ciborum et appetitus*. This is an extremely rare text, the only copy of which could be found at Naples—probably evidence of Huyberts's connections with Italian scholars. He continued his studies by enrolling as a student of law at Strasbourg in May 1653, and in the same year at Basel. At this point he was on his way to Italy, as in December he enrolled at Padua, while Tommaso Cornelio (1614–1684) later reported to have communicated his researches on nutrition, amongst others, to Huyberts and Rasmus Bartholin (1625–1698, who stayed in Italy in 1653–1656), when they came to Naples. In the second half of the 1650s he was again in the Netherlands, most probably at Leiden, and working on the figures (as reported by Clerselier) and on the editing of Descartes's *L'homme*, in an edition planned by the Elzeviers and announced in 1656 (as I discuss in Sect. 1.6). Such activity started therefore no later than 1656 and continued—through anatomical observations—until 1661–1662, as reported by Borch (as I discuss below). Moreover, Huyberts's observations probably continued in a more public form in 1662, when, in March, he enrolled again at Leiden even though he was already a *doctor medicinae*, which suggests he might have continued his anatomical activities within the University. Moreover, Huyberts, as reported to Niels Steensen (1638–1686) by some undisclosed correspondent of his, had discovered the vitelline duct before Steensen could announce its discovery in 1664, while in 1663–1664 Huyberts collected botanical specimens for Jacob Breyne (1637–1697). In October 1666, in turn, together with the Japanese-German-Dutch polymath Peter Hartzing (1637–1680), a friend of his, Huyberts was in charge of the reactivation of some mines by the Duke Johann Friedrich of Brunswick-Lüneburg (1625–1679) at St. Andreasberg (Harz). However, the enterprise ceased the following year, as a consequence of the Duke's demanding expectations on this enterprise. At that point, Huyberts probably came back to the Netherlands, buying at the beginning of 1670 a prebendary of the Oud-Munsterkerk in Utrecht, while later in the same year he undertook another voyage to Italy, where he came in contact with the Neapolitan physician Lucantonio Porzio (1639–1723) and with Adrien Auzout (1622–1691). In 1685 he married Lijdia Langlij (1642–after 1685), widow of Christiaan Melder (1625–1681, lecturer in mathematics at Leiden), and died in Utrecht in early 1716. During his life he was also acquainted with Johannes Swammerdam

---

[89] Descartes 1664b, *Préface*, 6 (unnumbered).

(1637–1680), Melchisédech Thévenot (c. 1620–1692), Matthaeus Sladus (1628–1689), Pieter Guenellon (c. 1647–1722), John Locke (1632–1704), Burchard de Volder (1643–1708), and Johannes Hudde (1628–1704).[90]

He was acquainted with Descartes's *L'homme* at least since his first studies at Leiden (1649–1652), whether or not by owning a copy: his *Disputatio de affectibus ventriculi*, concerning the process of dissolution of food and its transformation into chyle, bears witness to a textual reliance on it (as evident from Table A.1 in the Appendix), with regard especially to the topic of hunger (which was appropriated also by others, as mentioned above).[91] We can thus suppose that Huyberts got his copy (*Huyberts1*) around 1652, and in any case no later than 1656 (about which the Elzeviers informed Clerselier in 1657): in fact, in the same span of years when De Raey also most probably got a copy (as seen in Sect. 1.4.3). The two copies were thus probably related, regardless of their origin. Huyberts might have got it from De Raey, though we cannot exclude as possible sources other professors teaching at Leiden during Huyberts's studies, as Heereboord (who owned a copy) and Vorstius (who knew about the treatise, though there is no evidence that he had a copy), and, outside academia, Cornelis van Hogelande (1590–1662), whose *Historia oeconomiae corporis animalis* (1646) was probably used by Huyberts in his *Disputatio*, and who was a close associate of Descartes and the keeper of his trunk at Leiden in 1649–1650 (though we have no evidence that he was in possession of a copy).[92] However, De Raey (teacher at Leiden since 1647) is a more likely source (and, being his teacher, we can presume that he allowed Huyberts to get a copy and not vice-versa): their close association is testified, for instance, by Huyberts acting as witness at the

---

[90] Cornelio 1663, 191; Steensen 1667, 50; Swammerdam to Thévenot, 30 October 1670, in Swammerdam 1975, 58; Breyne 1678, 22, 67, 140, and 176; Denyssen 1682, dedicatees's page; Sladus to Johann Georg Graevius (1632–1703), 22 November/2 December 1684, in Locke 1976–1989, volume 2, 655–657; Egbertus Veen (c. 1630–1709) to Locke, 15/25 July 1689, in Locke 1976–1989, volume 3, 650–652; Locke to Graevius, 25 April 1690, in Locke 1976–1989, volume 4, 63–64; Benjamin Furly (1636–1714) to Locke, 8/19 February 1704, in Locke 1976–1989, volume 8, 189–191; Guenellon to Locke, 29 August/9 September 1704, in Locke 1976–1989, volume 8, 386–388; Porzio 1704, 97–98; Le Clerc 1709, 393; Gatterer 1785–1792, volume 3, 230–231; Du Rieu 1875, 396, and 496; Knod 1897, 255; Molhuysen 1913–1924, volume 3, 64; Wackernagel 1962, 492; Prins 1936, 28; Poelhekke 1961, 334; Helk 1971, 175; Westera 2018, 245–252; Stock and Weichert 2020; Strazzoni 2020; Het Utrechts Archief, toegangsnummer 34-4, inventarisnummer U048a004, aktenummer 20; toegangsnummer 711, inventarisnummer 99, 472; toegangsnummer 711, inventarisnummer 129, 321; Erfgoed Leiden en Omstreken, toegangsnummer 1004, inventarisnummer 23 Y, 235v; Stadsarchief Amsterdam, toegangsnummer 5001, inventarisnummer 41, 270; toegangsnummer 728, inventarisnummer 37.

[91] See *supra*, nn. 27, 66, and 74.

[92] Huyberts shares with Van Hogelande the characterization of such a corrosive part of the blood as 'menstruum', a term used in chemistry to label corrosive substances (one can find it, for instance, in the 1628 edition of Sennert's *Institutiones medicinae*, first edition 1611): Sennert 1628, 1301–1302. Moreover, like Van Hogelande, Huyberts relates such a *menstruum* to the concoction of dry food: Huyberts 1652, thesis 1. It was Van Hogelande who first adopted this term to label such a component of the blood, whose discovery he traces to Descartes: Van Hogelande 1646, 49–50 and 74–75. Descartes presented the idea of a substance making digestion possible in his *Discours* (AT VI, 53), and developed it to a larger extent in his *L'homme* (see the text quoted in Table A.1). We can at least presume that Van Hogelande discussed the matter with Descartes himself.

1.4 Other Dutch Copies

baptism of De Raey's daughter Elisabeth on 14 May 1659.[93] Also, Huyberts was a close acquaintance of other students and associates of De Raey, namely Hartzing and Hudde.[94]

Leaving aside the question of the priority of one copy over the other, their connections are further testified to by the fact that De Raey was well aware of Huyberts's efforts to prepare the figures and an edition of *L'homme*. This emerges, again, from Borch: according to what De Raey reported to him in October 1662, Huyberts was the only one who could "properly publish that book" for which purpose he had dissected 100 brains in one year.[95] Such dissections took place around 1661 at least: according to a letter of Borch to Thomas Bartholin (1616–1680) of 23 July 1661, indeed, the "Cartesian Huyberts" was carrying on, at his house, anatomical observations on the "connections, origin and direction" of the "fibres" or filaments which had already been discussed by Descartes, and was trying to show that they have the conarion or pineal gland as their centre.[96] To sum up, Huyberts was trying to shed light, by his brain dissections, on the nature of the features central to De Raey's criticisms of Schuyl's edition. Let us assume that *Huyberts1* and *DeRaey* thus shared the same deficiencies; but even in the case that their text was no less clear than those used by Schuyl and Clerselier, anatomical observations would have been required to understand, or to corroborate Descartes's neurophysiology. In fact, even if Huyberts at some point had access to Clerselier's text via *Huyberts2* and supposing that he trusted it (as in fact not only his associate De Raey, but also Hudde had a strong anti-French stance),[97] he still had to perform anatomical observations in order to complete his work, at least to prepare the figures. This can explain how De Raey, too, who might have had access to the same copy (and supposing that he trusted it), nonetheless criticized Schuyl's translation. Such difficulties in making sense of the text can partially explain why Huyberts never helped Clerselier with his edition: we know that he certainly recovered from a supposed condition of illness after 1657, though he never contacted Clerselier again. So he probably did not find his copy (*Huyberts2*) particularly useful, or—like his associates—just did not want to help Clerselier to appropriate Descartes's legacy. Eventually, the appearance of Schuyl's and Clerselier's editions probably put an end to Huyberts's plans.

---

[93] Erfgoed Leiden en Omstreken, toegangsnummer 1004, inventarisnummer 222, 172r.

[94] On Hudde, who during his studies at Leiden (from 1654) lived at De Raey's house, see De Jong 2018. On Hartzing and De Raey see Strazzoni 2020.

[95] "Adjecit neminem posse librum illum bene edere, excepto Huberto, qui vel 100 cerebra unico anno examinavit in ejus rei gratiam," Borch 1983, volume 2, 217.

[96] "D. D. Hubertus Cartesianus varia molitur, sed adhuc intra domesticos parietes, praesertim in cerebro occupatur expendendo, fibras ibi Cartesii persequitur, illarumque nexum, exortum, progressum, totus in eo, ut ad conarion tanquam centrum omnes derivet," Borch to Bartholin, 23 July 1661, in Bartholin 1663–1667, volume 2, 404; see also 374–375: Borch to Bartholin, 20 March 1661, according to which De Raey reported to Borch that Huyberts had unveiled Lodewijk de Bils's (c. 1624–1671) supposed secret techniques of embalming.

[97] See *supra*, n. 63.

1.5 Clerselier, Chevreau, and Andreae

As mentioned above, Clerselier claimed to be in possession of an autograph or "original" of Descartes's *L'homme*. A claim which (as I show in what follows) we have no sound reason to doubt. Yet it has raised some doubts among historians, in particular, as he once claimed to be in possession of a Cartesian manuscript, viz. a letter attributed to Descartes which he read at the Académie de Montmor on 13 July 1658, before declaring his forgery (which had been suspected by some when it was read) in the third volume (1667) of his edition of Descartes's correspondence. A letter which was textually based on some passages of Descartes's *Le monde ou le Traité de la lumière* (posthumously published in 1664). So Clerselier might have been feigning to be in possession of an autograph of *L'homme*, as well as of an autograph figure of the nerve-muscle system which he claimed to have at his disposal, without declaring its origin (and certainly not present in the autograph of *L'homme*, as I discuss below).[98]

To play devil's advocate, Clerselier was in fact not very consistent on how he got his manuscript, and on its condition. After having subtly announced his forthcoming edition of *L'homme* in his preface to the first volume (1657) of his edition of Descartes's correspondence,[99] he made a call for illustrators in the preface to the second volume (1659)—for which he eventually sent copies to Van Gutschoven and La Forge (as mentioned in Sect. 1.4.2)—and reported to have received the treatise (without specifying whether it was an autograph) from the one "who was the custodian of all the goods of his [viz. Descartes's] mind,"[100] namely Chanut, who, technically speaking, did not come into possession of the treatise as the custodian or trustee of the papers which Descartes left at Stockholm (the inventory of which does not contain either *L'homme* or his *Le monde*),[101] but more as a collector of his scattered papers. Also, in his 1659 preface Clerselier reports to be in course of giving the treatise "all the best form which is possible," while "amongst other things" the figures were missing, and remarking that "it was to be wished for its [viz. *L'homme*'s] final perfection, that its author himself had been able to put the last hand on it."[102] In fact, in the 1657 preface and in the one to the third volume (1667) of his edition of Descartes's correspondence Clerselier remarked that, with regard to the letters, he had supplied those parts of the text which were scarcely readable in the manuscripts, and had softened some of Descartes's tone in the most polemical ones,[103] so that we can suppose that he did the same with Descartes's *L'homme*: as in the case that

---

[98] Descartes 1664b, *Préface*, 15 and 18 (unnumbered); see Belgioioso 2005; Bitbol-Hespériès 2019.
[99] Descartes 1657–1667, volume 1, *Préface*, 12 (unnumbered).
[100] Descartes 1657–1667, volume 2, *Préface*, 14 (unnumbered).
[101] AT X, 5–12.
[102] Descartes 1657–1667, volume 2, *Préface*, 14 (unnumbered).
[103] Descartes 1657–1667, volume 1, *Préface*, 7 (unnumbered); volume 3, *Préface*, 15 (unnumbered). See also the review of Clerselier's edition which appeared on 15 January 1665 in the *Journal des sçavans*: "M. Des Cartes avait laissé ce traité dans une si grande confusion, qu'il ne serait pas intelligible si M. Clercelier ne l'avait mis en ordre," Anonymous (Sallo?) 1665, 11.

1.5 Clerselier, Chevreau, and Andreae

he had a scarcely readable manuscript, or that the text—although readable—was nonetheless not entirely clear in its argument: the text alone, without the figures, is not enough.

In turn, in the preface to his 1664 edition Clerselier provides a partially different story, specifying to have at his disposal an autograph, which he had had at least since 1657, when the Elzeviers heard about it (as mentioned in Sect. 1.4.3), and which he deems more reliable than Schuyl's text(s). Indeed, Clerselier reports that after having heard from Pollot that Schuyl was in possession of some autograph figures by Descartes (viz. those mentioned in Sect. 1.2), he asked Schuyl for them as well as for those he had made for his edition. Schuyl provided him with such materials (certainly, copies of) as well as with a copy (*Clerselier1*, tracing to around 1660–1661) of the treatise, from which Clerselier inferred that Schuyl was working on a defective "copy" (though it is not clear whether *Clerselier1* was a copy of one of the manuscripts available to Schuyl, or the result of their collation), and produced defective figures.[104] Moreover, in his 1664 preface Clerselier omits to refer to any changes he made to the treatise, with the exception of its division into parts and articles (viz. paragraphs provided with numbered titles, sometimes including unnumbered and untitled paragraphs in themselves), having found it "all continuous, without any distinction of chapters or articles," and willing to make it uniform with Descartes's *Description du corps humain*, provided by Descartes with a similar partition and published along with his *L'homme* by Clerselier.[105]

---

[104] Descartes 1664b, *Préface*, 2–4 (unnumbered). It is not easy to date with precision when Clerselier asked Schuyl for the figures. In any case, this was most probably not long after the appearance of the call for illustrators (which probably came as at that point Clerselier had time to focus on the edition of *L'homme*, after having completed the first two volumes of Descartes's correspondence), provided in the preface of the second volume of Descartes's correspondence, finished being printed on 28 May 1659 (Descartes 1657–1667, volume 2, 565, unnumbered), at which point Schuyl was thus already in possession of the autograph figures, which he got from Pollot. Having received the materials, Clerselier formally complimented Schuyl's work, hoping, however, that Schuyl was not going to speed up his publication plans, and instead that he was going to postpone the publication of his edition until after Clerselier's, through which Schuyl could amend his. Such compliments were however interpreted by Schuyl as an encouragement to publish his edition, and apparently coming after he got *Schuyl* (before or in early 1659) and *VanSurck* (1659–1661, as I show in a moment) (Descartes 1664b, *Préface*, 4–5, unnumbered; cf. *supra*, n. 7). Clerselier thus received the materials from Schuyl at a time at which it was evident that the latter was at a stage of editing more advanced than Clerselier's; at once, Clerselier seemed to be confident that he could provide better figures: viz. at a time at which Van Gutschoven and/or La Forge were at least working on them—i.e. around summer 1660—if not after having received their figures, viz. in February 1661, when La Forge had already delivered all his figures, and Van Gutschoven at least some of them. At that point, Schuyl had certainly received *VanSurck*. For a reconstruction of the chronology of the preparation of the figures of Clerselier's edition, see Strazzoni 2023a.

[105] Descartes 1664b, *Préface*, 28 (unnumbered). In turn, Schuyl's edition is divided into numbered paragraphs (sometimes broken by unnumbered ones), albeit with a different numbering, and with no titles or division into parts. The fact that both Schuyl's edition and ATH 1444 come to match, time to time, the paragraph division of Clerselier's edition, and that the paragraph division of ATH 1444 (probably prepared before Schuyl's edition, and unknown to Clerselier) matches more that of Clerselier than of Schuyl (as evident from Table A.2 in the Appendix) suggests that Clerselier had at his disposal a manuscript already divided into paragraphs, which he then numbered and provided

Plainly, in 1664 Clerselier was trying to sell his edition concurrently with Schuyl's in circulation: accordingly, he no longer mentioned the defects of his manuscript, and reported to have an autograph: a claim which we have in fact no reason to doubt, as after all Clerselier declared to be willing to show it (together with the above-mentioned autograph figure of the nerve-muscle system, the origin of which is more obscure than that of the manuscript, but which seems to be authentic, as I discuss below) to anybody interested.[106] Also, Clerselier reports that the Elzeviers had heard that he owned an autograph, and this amounts to them being an injured party should Clerselier's fraud be revealed. In fact, the ambiguity concerning Chanut's access to *L'homme* can also be explained, as Chanut was a custodian of Descartes's papers *lato sensu* (and, as I show in Sect. 1.6, also his first editor), and Clerselier had no cogent reason to declare who was Chanut's source (as not even Schuyl, in fact, declared under what circumstances Van Surck and Pollot got their copies).

As to Clerselier's access to the autograph, its reconstruction is nonetheless complicated by the fact that he also probably had access to two copies (besides *Clerselier1*), namely the one which Chanut got with the help of Chevreau (*Chanut*, as mentioned in Sect. 1.2), and another from Tobias Andreae (1604–1676), which are worth exploring. The above-mentioned *memoire* of Chevreau, in which his copy is mentioned, reads as follows:

> The last time that Mr. Ch[anut] went to Sweden, he arrived very late in Stockholm, and on arriving he asked me to go and see him. (…) I told him that I had a *Traité de l'homme* by Mr. Descartes, which he had sought in vain with all the passion he had (…). He asked me more than once how this treasure could have fallen into my hands; and I replied, as it is true, that I had it from Monsieur de La Voyette, gentleman to the Queen, who had been page to the Prince of Orange, and who had it from Madame Princess Elisabeth, a famous pupil of Mr. Descartes. As soon as he had seen the first leaves, he begged me instantly to lend it to him, with the promise to send it back to me from Hamburg, where he would take care to have it copied with the utmost diligence. Being in Paris, he shared it with Mr. de Clerselier, his brother-in-law, who put this badly copied manuscript in order, [and] who afterwards shared it with others: and we can see the destiny of this manuscript in the Preface of this *Traité* which was printed, although there is no mention of me, nor of Mr. de La Voyette.[107]

As mentioned in Sect. 1.2, the meeting of Chanut and Chevreau took place in May 1653, after which Chanut moved out of Sweden to Germany (where he was in July), and, passing through Hamburg (from where he might have returned *Chevreau* to its owner) and Holland, arrived at Paris around July or August (before assuming in November his post as Ambassador at The Hague: a position which he kept until November 1655).[108] In fact, only the information tracing to Chanut's presence in

---

with titles whenever he saw them fit a division into articles. ATH 1444, notably, contains some internal titles, concerning the explanation of the senses, absent in the other versions, certainly the result of the translator or copyists' intervention: "Tactus (…) Gustus (…) Odoratus (…) Auditus (…) Visus," ATH 1444, 32, 38, 40, 43, and 46.

[106] Descartes 1664b, *Préface*, 3 and 18 (unnumbered).

[107] Chevreau 1697, 101–102.

[108] Chanut 1677, volume 3, 186–195; Meyer 1679, 89; Hanotaux et al. 1884–1969, volume 21, 58 and 115; Wrangel 1891, 25; Huygens to Dohna, 24 July 1653, in Huygens 1911–1917, volume 5, 180–181; Raymond 1999, 117 and 119.

1.5 Clerselier, Chevreau, and Andreae

Sweden was first hand: the other details reported by Chevreau were certainly developed on the basis of Clerselier's 1659 preface (viz. the reference to a badly copied manuscript and to its delivery to Clerselier by Chanut) and his 1664 one (as to Clerselier's sharing the text with others). Either Chevreau had simply forgotten that Clerselier had mentioned having an autograph at his disposal (in 1664) or he might be insinuating that Clerselier feigned to have it without, of course, acknowledging a source. In any case, it makes sense that in that year he did not mention any copy at his disposal, if he had an autograph.[109] In turn, it may appear surprising that Clerselier did not come into possession of *Chanut* (or a copy of it) at least until July 1654: at which point, indeed, he was still trying to obtain a copy from Andreae, as emerges from their extant correspondence, tracing from around the beginning of 1654 to 16 July of the same year.

To sum up its contents, around the first half of 1654 Andreae wrote a now lost letter (letter A) to Clerselier reporting to have at his disposal some Cartesian manuscripts (as will be evident in what follows). To this letter, Clerselier replied by sending him, through one of the sons of Andreae's father-in-law Lodewijk de Geer (1587–1652; as revealed by a further letter, C, which I discuss in the next paragraph), a letter surviving as an undated draft (letter B), providing him with a copy of the so-called 'Stockholm inventory', i.e. the list of the papers that Descartes left at Stockholm when he died (and prepared by Chanut and probably Christiaan Huygens, 1629–1695, between the end of 1653 and the beginning of 1654).[110] In letter B, Clerselier (1) maintains not to have at his disposal those items which are unmarked in the attached inventory, for the reason that Chanut enclosed them in some package or with some other item (the text is defective) which had not been sent to France.[111] Moreover, he (2) asks Andreae, who, as reported by Clerselier, wrote him (viz. by letter A) to be in possession of a certain treatise *De homine*, "begun, not finished," to check whether the beginning of such a treatise was the same as the first two articles of Descartes's *Description*, of which he was in possession of the autograph and of which he provides a partial transcription in the letter. Clerselier asks Andreae, in the case that the texts do not match, to make a copy of it and send it to him.[112]

Afterwards, Clerselier received from Andreae a now lost copy of letter A, as revealed by his letter of 12 July 1654, i.e. letter C, reacting to it.[113] According to Clerselier's letter C, letter B was not lost, but just not yet delivered to Andreae (as he heard from Johannes Clauberg, 1622–1665, that De Geer, to whom he gave letter B, was coming back from his voyages). Accordingly, in letter C, Clerselier just provides a summary of letter B (but without mentioning whether he had received all

---

[109] Meschini 2011, 194.

[110] As reconstructed in Descartes 2003, xvi–xxi.

[111] Adrien Baillet (1649–1706) provides an account of the delivery of the manuscripts to Paris, without however providing a source, and wrongly claiming that they arrived in 1653: Baillet 1691, volume 2, 402 and 428. See Descartes 2003, xvi–xxi.

[112] Clerselier to Andreae, first half of 1654, in AT X, 13–14; commented on in Descartes 2003, xxi–xxii.

[113] Published and commented on in Dibon 1990: see esp. 497–499.

the remaining papers from Chanut), adding at the same time some details: besides remarking his having provided in his previous letter (B) a transcription of the first two articles of his *Description*, he asks Andreae to send him copies of Descartes's letters to Andreae himself, to Henry More (1614–1687), and "others," but those to Elisabeth and Chanut (all evidently reported in letter A, being those to Chanut and "others" briefly mentioned also in letter B) were not, apparently, needed by Clerselier, who already had good exemplars of them, either originals or copies.[114]

As evident from this correspondence, Clerselier was not yet in possession of a copy of *L'homme* as late as July 1654: at which point he repeatedly asked Andreae to exclude that his manuscript was that of Descartes's *Description*, apparently the only text which was in possession between the two. Probably, Clerselier had been enticed by Andreae's description of the item, labelled as "begun, not finished." Indeed, Clerselier must have noticed that Descartes's *Description* was evidently incomplete, and also showed a change of focus during its preparation.[115] In any case, most probably Andreae had at his disposal a copy (*Andreae*) of *L'homme* rather than of the *Description*. Indeed, it is unlikely that Descartes would have shared with him an incomplete treatise, in fact a failed project born out of the attempt at re-writing *L'homme* (which might have also resulted in the insertion of author's variants, around 1647, in the autograph in possession of Clerselier, whose edition shows notable differences with respect to Schuyl's and to ATH 1444), as he himself admitted not to have been able to complete the treatise and to provide a full account of animal generation.[116] In turn, Andreae could have labelled *L'homme* incomplete given the

---

[114] The origin of such Cartesian materials in Andreae's hands is not clear. He was well connected with Cartesians such as De Raey and Clauberg (see Agostini 2009, volume 1, 142–145; Strazzoni 2022): Clauberg was an intermediary of Andreae and Clerselier, therefore, he was not the source of Andreae's materials, as otherwise Clerselier would have asked for them from Clauberg himself. Moreover, as reported by Borch in his diary on 29 July 1661, Descartes lived for long at the house of Andreae's father-in-law De Geer at Amsterdam (Borch 1983, volume 1, 185). As reported in the letters of Descartes to him (in Baillet's paraphrases), they met once before 27 May 1644, while afterwards Andreae helped him during the Groningen *affaire* (Descartes to Andreae, 27 May 1644, in AT IV, 123), namely Descartes's quarrel with Martin Schoock (1614–1669). Their relations went beyond this, since in 1645 Descartes kept Andreae informed about his anatomical observations on plants and animals (Descartes to Andreae, 16 July 1645, in AT IV, 245–247): thus it is plausible that Descartes could have at some point shared with him his *L'homme* (along with other *Cartesiana*).

[115] The *Description* consists of five parts, namely (1) a preface with a summary of the contents of the treatise, (2) a discussion of heart-beat and blood circulation, (3) nutrition, (4) generation, and (5) growth. Notably, in the preface of the treatise there is no mention of the two latter processes, as Descartes lists only the physiological functions already dealt with in his *L'homme* (heart-beat, blood circulation, nutrition, the generation of animal spirits, sense perception, memory, and muscular movement), but without treating them all: AT XI, 226–227. See *infra*, n. 116.

[116] The preparation of Descartes's *Description* developed as an attempt to re-write his *L'homme* and to provide an account of animal generation and growth, following the genetic approach of his physics, for which, while writing *L'homme*, he lacked the necessary knowledge (as mentioned in his *Discours*: AT VI, 45–46). After the publication of his *Discours*, in February 1639 Descartes deemed such a goal feasible in the event he was to re-write his *Le monde* (including in fact his *L'homme*; Descartes to Mersenne, 20 February 1639, in AT II, 510–519), while deeming his *L'homme* as a just drafted or incomplete treatise (Descartes to Elisabeth, 6 October 1645, in AT IV, 309–310; Descartes to William Cavendish (1592–1676), October 1645, in AT IV, 325–330). Eventually, the explosion

fact that it refers to the subsequent treatment of the rational soul, which is also mentioned in Descartes's *Discours de la méthode* (1637), but which is absent from all the versions of *L'homme*,[117] or, having been informed of Descartes's anatomical activities,[118] as well as of his opinions on his own treatise, he just deemed it unfinished in the light of Descartes's overall aim of providing an account of animal generation.

In any case, we can explain why Chanut had not yet shared the copy *Chanut* with Clerselier as late as in July 1654: indeed, at least until January 1654 it was Chanut, not Clerselier, who was planning to publish Descartes's manuscripts: at least, a part of his correspondence.[119] Only afterwards did the editorial project (and the manuscripts in

---

of the *affaire* Regius prompted Descartes to prepare a new treatise in physiology, as during it (1646–1648) he repeatedly distanced himself from Regius's appropriation of his *L'homme* by labelling it (and its copies) as—again—imperfect, as in the letters mentioned *supra*, n. 26. In particular, on 31 January 1648 he wrote to Elisabeth that since his *L'homme* had been badly transcribed by many, he decided to put it in order (*au net*), viz. to re-make it. Such a new treatise was in any case following the same purpose of his *L'homme*, namely to present the functions of an already formed animal. In the course of the preparation of such a treatise, however, he reverted to his aim of providing a treatment of the generation and growth of the animal, on which he had ventured himself in the previous 8–10 days (Descartes to Elisabeth, 31 January 1648, in AT V, 111–114). This, plainly, matches the change in contents of his *Description* (see *supra*, n. 115), and also his dealing with the issue of generation in February of the same year, in his *Primae cogitationes de generatione animalium* (as reconstructed in Descartes 2000, *Introduction*). The same story is also provided in a letter to an undisclosed recipient of 8 April 1648, where Descartes reports to have long since abandoned his project of putting in order his description of the functions of the animal, as while reflecting on this subject (apparently, while re-writing his *L'homme* in the form of the *Description* between 1647 and 1648), he had made some discovery which would enable the completion of his physics viz. to treat the topic of generation (Descartes to unknown, 8 April 1648, in AT V, 259–261; see Bos 2004). This plan, however, soon came to a halt, as in the 'Conversation with Burman', which took place at Egmond-Binnen on 16 April 1648 with Frans Burman (1628–1679), Descartes, after having reported the same story of his reverting to a genetic approach while he was working on a treatise on the animal (i.e. the *Description*) during the winter, admits to have found this approach too long to be followed, and interrupted its writing, an interruption which therefore took place between 8 and 16 April 1648 (AT V, 170–171). On the reasons why Descartes did not publish his *L'homme* see Cook 2021.

[117] Descartes 1662, 35; Descartes 1664b, 29; ATH 1444, 33; AT VI, 59. This was eventually prepared, as a completion of Descartes's project, by La Forge, who in 1666 published his *Traité de l'esprit de l'homme*; on La Forge's plan, see Descartes 1664b, *Preface*, 12 (unnumbered); see also La Forge 1974, *Introduction*.

[118] See *supra*, n. 114.

[119] This is evident in from the correspondence of Constantijn Huygens and others of December 1653–January 1654, discussed in Descartes 2003, xvii–xxi: see Huygens to Elisabeth, 31 December 1653, in Huygens 1911–1917, volume 5, 194 (reporting that Chanut wanted to go through Descartes's papers with Christiaan Huygens in order to check whether there was still something to be printed of mathematical or philosophical importance, within or besides a selection of letters Chanut aimed at publishing); Johan de Witt (1625–1672) to Colvius, 3 January 1654, in Thijssen-Schoute 1967, 87 (according to which Chanut had already started the publication process of the letters, deemed as already in course of printing, while Descartes's manuscripts, according to him, were fragmentary and had to be re-ordered in the event of publication); De Witt to Colvius, 16 January 1654, in Thijssen-Schoute 1967, 88–89 (where De Witt claims to have understood that Chanut wanted to publish only those letters clearly disposed for publication by Descartes, but not other letters or works). See also *infra*, n. 134.

Chanut's possession) pass to Clerselier, who therefore wrote letter A not before 1654, when he started to be the leading figure of the Cartesian editorial project, and who at the time of letter B (viz. in the first half of 1654) had however not yet received all the manuscripts from Chanut, but only a part, as revealed by such a letter. Eventually, the manuscripts mentioned in the correspondence of Andreae and Clerselier were probably delivered to the latter before or during December 1654 (as revealed by Clerselier's access to some of the materials mentioned in his correspondence with Andreae).[120] Whether or not these included a manuscript of *L'homme*, this was certainly a copy (hypothetically, *Clerselier2*, copy of *Andreae*): indeed, as seen above, Clerselier asked for a copy of the treatise (explicitly in letter B, and showing interest also in letter C), and mentioned only Chanut as the source of the autograph.

Chanut, in turn, certainly secured the autograph between mid-1653 and late 1655, when he was in the Netherlands as Ambassador. The autograph he obtained was most probably *Descartes*: provided, however, with several additions and variants with respect to the versions published by Schuyl and to ATH 1444. While some of them could be translation variants, others consist of entirely new passages, which one can find only in Clerselier's edition (as evident from the edition of ATH 1444 provided in this book). So that they might be traced to Descartes's intervention on the text after the copying of *VanSurck*, and/or to Clerselier (as suggested by his 1659 preface). If at least some of them are by Descartes, he most probably inserted them in the text around 1647, namely when he aimed at putting his treatise "in order," this resulting in the attempt of re-making in the form of his *Description*.[121] Alternatively, we might think this re-writing resulted in a second autograph, which, however, should

---

[120] First of all, there is evidence that Clerselier received the letters of Descartes to Andreae: even if Clerselier did not publish the letters of Descartes to Andreae in his 1657–1667 edition, Baillet reports some extracts from Descartes's letters to Andreae, which were collected by Jean-Baptiste Legrand (d. c. 1704) after Clerselier's death. As argued by Bos, it might be that Legrand got such letters from Clerselier's heirs, as it was for other letters already in the possession of Clerselier (Baillet 1691, volume 1, xxii; Descartes and Regius 2002, xxiii–xxiv.) Moreover, Clerselier published, in his edition of Descartes's correspondence, the French translation of a complete version of an undated letter of Descartes to (probably) a certain Petrus van Buitendijck, published in a partial, original Latin form for the first time in 1653 by Andreae in his *Methodi Cartesianae assertio*: cf. Andreae 1653a, 947–949, with Descartes 1657–1667, volume 2, 53–55; re-used in Revius 1654, part 1, 364–367; Clauberg 1655, 355–360. Such a letter might have been among those to "others" mentioned in letter B and letter C. Clerselier probably received the materials before December 1654: indeed, on 12 December 1654 he wrote to More asking for the letters of Descartes he thought not to own and to be in More's possession, even if in fact at that point Clerselier was already in possession of their complete correspondence (Clerselier to More, 12 December 1654, in Descartes 1657–1667, volume 1, 308–310; More to Clerselier, 14 May 1655, in Descartes 1657–1667, volume 1, 310–315). Andreae might have delivered the materials to Clerselier after the latter contacted More: but it is more probable that he turned to More after having assessed that Andreae could not help him i.e. after having received materials from him: afterwards, he was forced to turn to More. More's letter to Descartes of 11 December 1648 and Descartes's letter to More of 5 February 1649 were used by Clauberg (Clauberg 1652, 4 and 474–475) and, concerning the letter of 5 February only, Andreae (Andreae 1653b, 280–281) and Christoph Wittich (1625–1687; see Wittich 1653, 154–155), though, as argued *supra*, n. 114, such materials most probably came from Andreae, not from Clauberg.

[121] See *supra*, n. 116.

have been deemed as not in need of completion by Clerselier in his 1659 preface, as seen above.

If we want to venture some hypothesis on Chanut's access to the autograph, an obvious source was Descartes's Leiden trunk. Among the witnesses of its opening, we can exclude De Raey (who evidently never wanted to help the French) and La Voyette (who got a copy from Elisabeth). Also Frans van Schooten jr. (1615–1660) is a quite weak candidate: indeed, Daniel Elzevier had asked Chanut about a number of texts including *L'homme* already in 1651–1653, at which time Van Schooten had already worked with the Elzeviers' establishment, viz. in preparing the figures for the first edition of Descartes's *Principia* (1644):[122] so that he would have probably provided them with a copy, if they were interested. Also, there are no copies of Descartes's *L'homme* (which he would have probably retained) among his manuscripts, now extant at Groningen, in which is present a copy of Descartes's *Compendium musicae*.[123] Leaving aside Van Hogelande—we have no direct evidence that he had a manuscript—the last remaining candidate is Van Surck, who in fact had kept Cartesian manuscripts after Descartes's death, which are however now lost.[124]

Eventually, it is worth taking some trouble over the above-mentioned autograph figure of the nerve-muscle system in the possession of Clerselier, and also over a diagram of sounds published by him—both being the items in possession of Schuyl as autographs, as touched upon in Sect. 1.2. It is unclear when and where Clerselier got the figure of the nerve-muscle system, labelled by him as a rough sketch (as Schuyl did with regard to his one).[125] As he was looking for figures in 1657, at that point he probably already had it in his hands (e.g. along with Descartes's papers), though he might have deemed it useful only at a later stage, when editing the text for his publication. We can in any case exclude that it came to Clerselier together with the autograph of *L'homme* not only because he himself deemed it figureless (as mentioned above), but also because, in that case, he would probably have shared the figure with Van Gutschoven and La Forge, who however provided figures different from the one hypothetically present in Descartes's autograph, included by Clerselier, as an "extract," in his edition (cf. Figs. A.2, A.3, and A.4 in the Appendix).[126] If Van Gutschoven provided a representation of the tubes connecting the muscles in an X-shaped disposition (different from the one by La Forge), this might have been inspired by the figures provided by Schuyl—including the autograph published by him (cf. Figs. A.1 and A.5 in the Appendix), as Van Gutschoven's drawings were probably finished after the appearance of Schuyl's 1662 edition.[127]

Notably, Clerselier also included in his edition a diagram of sounds different from the one provided by Schuyl (cf. Figs. A.6 and A.7 in the Appendix), who reported to

---

[122] See *infra*, n. 141.

[123] Groningen, Universiteitsbibliotheek, ms. GN108, 60r–83v.

[124] Descartes 2003, xv. See *supra*, n. 92.

[125] Descartes 1662, *Ad lectorem*, 32 (unnumbered; see *supra*, n. 7); Descartes 1664b, *Préface*, 15 and 18 (unnumbered).

[126] Descartes 1664b, *Préface*, 18 (unnumbered).

[127] For a reconstruction, see Strazzoni 2023a.

have a roughly sketched autograph of it.[128] It is unclear who prepared the diagram for Clerselier's edition, who is silent about it. In fact, both the versions of the diagram are incorrect, in particular the one published by Schuyl (Fig. A.6).[129] In it, lines D and E lack a proper division (as A and D, and A and E should be in the ratios of 3:4 and 4:5 respectively), while line F should be divided into 9 parts (to be in a ratio of 8:9 with A, viz. the ratio expressing a major whole tone), as indicated, in fact, in the *errata* appended at the end of the edition (both in the 1662 and 1664 publications of it).[130] Moreover, line A appears to have been divided, at first, into 4 parts (correctly holding a ratio of 1:2 with B, divided into 8 parts), and then—by dotted vertical lines—into 8 (correctly holding a ratio of 2:3 with C, divided into 12 parts), showing an ad hoc modification. Accordingly, if (1) the ad hoc, additional division is present in the diagram itself, which is (2) defective as to lines D and E, but not corrected at all, and if (3) the diagram is corrected in the *errata* as to line F, we can interpret the (1) ad hoc division as present in the original manuscript, and (3) the *errata* correction as by Schuyl, who, however, (2) did not indicate corrections for lines D and E, even outside the diagram. This might be explained by considering that the text edited by him (unlike the ones of Clerselier and ATH 1444) awkwardly refers, as to line F, to a "major sound" instead of a "major tone," so that he probably used French manuscripts all containing 'son' instead of 'ton', and inserted a correction, only in the *errata*, exactly addressing this point, while lines D and E could be corrected or figured out as divided into certain parts, by a potential reader, on the basis of the text alone.[131] The version published by Clerselier, on the other hand, is correct in all its lines but F, which holds with line A the ratio of 5:6, expressing a minor third, not a major whole tone or major second. In fact, there is no indication that Clerselier, too, might have derived the diagram from a manuscript: in principle, the mistake present in it can be attributed to someone preparing the diagram on the basis of the text only (e.g. Clerselier or someone on his behalf), and not necessarily according to the tradition of an incorrect manuscript of the diagram. Alternatively, we can suppose that this figure, like the one of the nerve-muscle system, also had a manuscript circulation before being published by Schuyl and Clerselier (even if both the figures are absent from ATH 1444).

---

[128] See *supra*, n. 7.

[129] For a correct version of the diagram, see Descartes 2011, 23 and 171–172.

[130] Descartes 1662, 122 (unnumbered).

[131] Cf. Descartes 1662, 42–43; Descartes 1664b, 36–37; ATH 1444, 44–45; signalled in the present edition. The diagrams were not modified in any of the seventeenth-century editions of the treatise (on them, see Van Otegem 2002, chapter 9), but in a Dutch edition of 1692, in which all the lines but A are equally, and wrongly, divided into 5 irregular parts: Descartes 1692, 246.

1.6 The Elzeviers' Editorial Plans

Before turning to a discussion of ATH 1444, it is also worth considering the Elzeviers' attempts at publishing *L'homme*, which intersect with Chanut's access to it, and with Clerselier's edition. As seen in Sect. 1.4.3, in early 1657 Louis and Daniel Elzevier asked Clerselier for a copy of *L'homme* to be sent to Huyberts, who had worked on the figures, in order to check whether he had done good work and to resolve some difficulty related to it. In fact, in the preface to their 1656 edition of Descartes's *Principia*, part of their 1654–1656 edition of Descartes's *Opera philosophica* (finished being printed before or during November 1656),[132] the Elzeviers announced that "if we will find [it] acceptable, we are willing to publish in a short while also Descartes's *Tractatus de homine* with genuine figures (hope of which has already been given us by a friend)."[133] This 'friend' was most probably Huyberts himself, on whose work the Elzeviers had reservations, as evident both from their request to Clerselier and from this announcement. In fact, the Elzeviers (especially Daniel) had been interested in the treatise since the early 1650s. This is revealed by Daniel Lipstorp (1631–1684) in his *Specimina philosophiae Cartesianae* (1653), reporting to be expecting the publication of Descartes's *L'homme* and *Description*, as well as of his letters, his "*Compendium of Mechanics* to be obtained from [variously] disseminated papers of others" (most probably Christiaan Huygens) and writings on analytic geometry: all texts about which Chanut had "given hope" first to the mathematician Johann Adolf Tassius (1585–1654), whom he met in Hamburg while returning from Sweden, and then to Daniel Elzevier, whom Chanut met at Lübeck, and who "had urged" him about them.[134] The meeting with Tassius took

---

[132] As evident from Van Schooten to Christiaan Huygens, 11 February 1656, in Huygens 1888–1950, volume 1, 381–382; Van Schooten to Huygens, 20 November 1656, in Huygens 1888–1950, volume 1, 512–513.

[133] Descartes 1656, *Typographus ad lectorem* (unnumbered).

[134] "(…) ardenter expectamus copiam illarum rerum, quas adhuc asservari novimus tum apud alios, tum imprimis apud (…) Chanutum, (…) antehac Legatum ad Augustissimam Sueciae Reginam, nunc ad tractatus pacis Lubecae (…). In iis autem volente Deo habebimus Tractatum de homine, cuius meminit philosophus noster Princip. philos. parte 4, art. CLXXXVIII, habebimus volente Deo epistolas magno numero a Nob. Cartesio ad amicos scriptas, rerum philosophicarum plenissimas, Compendium item mechanicarum diffusis aliorum libris comparandum, et dubio procul compluria analysin geometricam spectantia. Et iam spem earum fecit Vir Illustrissimus antehac Clarissimo Dn. Iohanni Adolpho Tassio, Hamburgensium Mathematico, cum Hamburgi esset ex Suecia rediens, tum nuperrime Dn. Danieli Elzevirio, qui ipsum Lubecae compellaverat. Ita ut quo tanto securius huius Illustris Viri liberalitate frui possimus, id unicum nobis incumbat, ut pro eius incolumitate preces fundamus. (…) [S]i Deus Illustrissimo Viro Dn. Petro Chanuto (…) vitam concesserit (quod speramus, et ardenter optamus) videbimus Tractatum de homine, Tractatum de generatione animalium, Epistolas magno numero ab authore ad amicos scriptas, et ab is receptas, rerum philosophicarum plenissimas, Compendium mechanicarum diffusis aliorum libris comparandum, et alia Analysin geometricam spectantia quam plurima," Lipstorp 1653, 25 and 84. Notably, it appears that Chanut had already separated himself from Descartes's treatise on mechanics i.e. his very short *Explication des machines et engins*, which is mentioned in the Stockholm inventory, but was not delivered to Clerselier and which was at some point in the hands of Christiaan Huygens: see Huygens's commentary on Baillet 1691, volume 1, 268: Huygens 1888–1950, volume 10, 402; Van Otegem, volume

place in June or July 1651, after Chanut had concluded his ambassadorial service in Sweden,[135] while the meeting with Elzevier at Lübeck, where Chanut stayed for long periods from June or July 1651 to April 1653 as a mediator during the peace talks of the Second Northern War,[136] took place no later than March 1653.[137] Interestingly, Descartes's *L'homme*—definitely not in Chanut's hands in early 1653—is not overtly labelled as to be retraced from others' papers, as is, on the other hand, Descartes's *Compendium of mechanics* i.e. his *Explication des machines et engins*, about which, however, Lipstorp might have been informed by Huygens himself, an acquaintance and correspondent of his, more precisely than about *L'homme*.[138]

From this it emerges not only that Chanut had to be the first editor of Descartes's manuscripts—at least his letters—as mentioned in Sect. 1.5,[139] but also that this—at least in their expectations—had to happen with the Elzeviers as publishers. Such a potential cooperation in any case came to an end in 1654, when Clerselier became the main editor of the writings. At that point, the Elzeviers' establishment was to publish the third Latin edition of Descartes's works (1654–1656; in fact the second, after the one of 1650), i.e. his *Opera philosophica, editio tertia*, including his *Meditationes* and other texts, such as the *Epistola ad Voetium* (1654), and his *Principia, Specimina philosophiae* (including Descartes's *Dissertatio de methodo, Dioptrica*, and *Meteora*), and *Passiones animae* (1656).[140] The Dutch Cartesians were involved in the edition by the Elzeviers: De Raey took care of revising the text of the *Principia* and *De methodo*, while Van Schooten that of the *Dioptrica* (for which he also

---

2, 554–555. So that Chanut had already given such a piece to (most probably) Huygens before he asked his help in sorting out Descartes's manuscripts between 1653 and 1654; this confirms that Chanut did not want to publish all Descartes's papers: see *supra*, n. 119. On Descartes's posthumous mathematical manuscripts (including analytic geometry), see Di Loreto 1995.

[135] Chanut 1677, volume 2, 229–233; Neumann 1935, 126.

[136] Wrangel 1891, 25.

[137] Indeed, Lipstorp's *Specimina* was presumably completed before 1 April of the same year, viz. the date of its dedicatory letter: Lipstorp 1653, *Principi Johanni*, 8 (unnumbered). The *Specimina* might have been the source of Vopiscus Fortunatus Plempius (1601–1671), who in the third edition (1653) of his *De fundamentis medicinae* (1638) reports to have heard that Descartes's *L'homme* (which he did not have access to, and which he discusses with regard to the pineal gland) was in course of printing: Plempius 1653, 116. This edition of Plempius's book was published after 6 October 1653, or the date of the *imprimatur*: Plempius 1653, 388 (unnumbered).

[138] Their extant correspondence (five letters) is in Huygens 1888–1950, volume 1.

[139] See *supra*, n. 119.

[140] A so-called second edition, appearing in 1650, included the same treatises. The first edition consisted of the publication of Descartes's *Principia* and *Specimina* in 1644 (though never sold as a first edition of Descartes's *Opera philosophica*): see Van Otegem 2002, volume 2, 680–683.

provided new figures and notes),[141] after a request for corrections from Louis Elzevier dating to circa October 1654.[142] We do not know whether Huyberts had been involved in the project yet: probably, this happened only afterwards i.e. in 1656, as even Descartes's *Géométrie*, edited and translated by Van Schooten for its 1649 Latin edition (by Johannes Maire, 1603–1657),[143] and whose woodcuts were hence bought, probably in 1654, by Louis Elzevier from Maire along with those of the *Dioptrique* and *Météores* (1637),[144] was not clearly planned for inclusion in Descartes's *Opera* in 1654, being announced as forthcoming in some weeks in the Elzeviers' 1656 preface announcing also the edition of *L'homme*, but eventually published only in 1659 (edited by Van Schooten).[145] So that in 1654 the main aim of Louis Elzevier appears to have been that of re-publishing those texts which appeared also in the 1650 edition of the *Opera*,[146] after which other texts—such as the *Géométrie* and *L'homme*—had to appear. At that point (1656), the autograph of *L'homme* was already in the hands of Chanut or, more probably, Clerselier: so that the Elzeviers were forced to revert to the latter in early 1657. Later, Huyberts's apparent difficulties in making sense of the text (and illustrating it), as well as the editorial enterprises of Schuyl and Clerselier somehow obscured the Elzeviers' plans with regard to *L'homme*, and probably their original plan of publishing also Descartes's letters, about which Daniel Elzevier urged Chanut, as seen above. Nonetheless, in 1668 Daniel Elzevier published a Latin translation of the first two volumes of Clerselier's edition of Descartes's correspondence, though with some variants with respect to Clerselier's text, so that most probably the edition was prepared with the help of some manuscript circulating in the Netherlands.[147] In the preface to the first volume, notably, Elzevier announced the

---

[141] Descartes 1656, *Typographus ad lectorem* (unnumbered). For such a text (and of all the *Essais*) Van Schooten had already prepared the drawings, and most probably the woodcuts, for its first edition (1637, published by Johannes Maire), and he did the same for Descartes's *Principia* (1644), while he did not prepare the woodcuts for the 1644 Elzevier edition of Descartes's *Specimina* (including the *Dioptrica*), for which new ones were used: for a complete reconstruction, see Descartes 2007, 64–66 and 72–73; Van Otegem 2002, volume 1, 49.

[142] As evident from Van Schooten to Huygens, 25 October 1654, in Huygens 1888–1950, volume 1, 299–301; Van Schooten to Huygens, 23 December 1654, in Huygens 1888–1950, volume 1, 312–313.

[143] Van Otegem 2002, volume 1, 103–107.

[144] In 1653 Van Schooten had contacted Maire, the first publisher of Descartes's *Dioptrique*, to publish a new Latin translation of it (by Van Gutschoven), together with an essay *De refractionibus* by Huygens; the latter was hence asked by Van Schooten, in his letter of 25 October 1654, to publish it in the new Elzevier edition of the *Dioptrica* (viz. what was to be the one of 1656), as Louis Elzevier had bought its woodcuts from Maire. See Van Schooten to Huygens, 5 June 1653, in Huygens 1888–1950, volume 1, 233–234; Van Schooten to Huygens, 25 October 1654, in Huygens 1888–1950, volume 1, 299–301; Dijksterhuis 2005, chapter 2.

[145] Descartes 1656, *Typographus ad lectorem* (unnumbered); Van Schooten to Huygens, 23 December 1654, in Huygens 1888–1950, volume 1, 312–313; Van Otegem 2002, volume 1, 120–122.

[146] See *supra*, n. 140.

[147] Descartes 2003, xxiii.

forthcoming publication of a Latin edition of *L'homme*, accompanied by Latin translations of La Forge's *Remarques* on it and of his *Traité de l'esprit de l'homme* (ideally completing Descartes's original plan).[148] However, only the Latin translation of La Forge's *Traité* was separately published by Elzevier in 1669 (as *Tractatus de mente humana*), while the announced Latin edition of Descartes's *L'homme* was published by Elzevier only in 1677, as a translation of Clerselier's edition of *L'homme, Description* and La Forge's *Remarques* (which, in turn, in the same year had a second edition, including Descartes's *Le monde*).[149]

1.7 ATH 1444

On the ground of the evidence on the manuscript circulation of Descartes's *L'homme* so far acquired, we can give attention to ATH 1444.

First of all, it is worth remarking that (1) the manuscript is certainly a copy of a further manuscript, not a first draft (as it is well ordered, and containing at least a peculiar omission, as I discuss below). Moreover, (2) the consulted sheets are figureless: though, as (A) the pineal gland is systematically indicated with 'FF' instead with 'H' (as in the other versions), (B) from time to time other reference letters for parts of figures show variants, and (C) a variant contains a remark about the different quality of the figures mentioned in the manuscript, it might be that some figures accompanied the text, or a manuscript from which it derived. At once, in the same sentence a part of the text is evidently missing: this includes a reference to another figure afterwards alluded to in the text (so that the omission was certainly a mistake), and makes the whole argument quite hard to understand, if not inconsistent (while its being at least grammatically consistent suggests that someone tried to correct the text).[150] So that certainly the illustrator, the translator, and the copyist were three different persons. (3) The consulted text is incomplete (viz. it is interrupted at

---

[148] Descartes 1668–1683, volume 1, *Typographi praefatiuncula ad lectorem*, 2 (unnumbered). See *supra*, n. 117.

[149] As hypothesized by Van Otegem, it might be that Elzevier asked Schuyl to take care of his edition (Descartes 1677a) during the 1660s: Schuyl's death (1669), however, prevented him to start or finish his work: Van Otegem 2002, volume 2, 507–511. Thus far, the identity of the 1677 Latin editor is unknown: we can reasonably exclude that De Raey could have worked on Clerselier's version, moreover, as his translation, as mentioned in Sect. 1.4.2, was labelled as "evidently new" in a 1723 catalogue (see *supra*, n. 70), and it was certainly different from all the published ones. The 1677 Latin text is moreover different from that of ATH 1444. Van Otegem has discussed the possibility that the French edition of the treatise published in 1680 at Amsterdam by Guillaume le Jeûne—in fact a pseudonym for Henricus Wetstein (1649–1726)—was carried out with materials originally belonging to Daniel Elzevier, who died on 13 October 1680; the matter is worth further exploration: Van Otegem 2002, 527.

[150] Cf. Descartes 1662, 98–99: "figura 50: et inter eundem dispositum ad videndum obiectum paulo propius situm, uti ex figura 51 apparet: quae non in eo tantum consistit, quod humor crystallinus est paulo incurvatior, aliaeque oculi partes"; Descartes 1664b, 87: "la 50. figure p. 65 et le même œil, disposé à en regarder un plus proche, comme il est en cette 51 qui consiste, non seulement en

what in Clerselier's edition is article 91 on 106, and the beginning of a fascicle is missing),[151] and (4) the origins of the manuscript itself are at present still obscure. As is well known, the Bibliotheca Thysiana at Leiden was established following the will of the book collector and member of a family of scholars Johannes Thysius (1622–1653).[152] However, I could not find mention of the manuscript in the early-modern catalogues related to the library.[153] In any case, even if Thysius enrolled as a student of arts and then law at Leiden in 1635 and 1648–1652 (when he might have come in contact with the treatise),[154] his library came to be expanded after his death, so that the manuscript could just have entered it afterwards. Also, (5) we still cannot identify the translator. The texts of those who appropriated to various extents Descartes's treatise, as Regius, Huyberts, and Guillaume Philippi (c. 1600–1665), do not offer evidence in this regard,[155] not to mention the Latin translation of Clerselier's edition, which appeared in 1677 and does not match the text of ATH 1444.

In any case, we can decidedly exclude De Raey as a translator, as, even if certain additions and underlinings concerning the structure of nerves points to an interest in the topic at least by the copyist,[156] (1) De Raey's criticisms of Schuyl's rendering of the structure of nerves apply also to the text of ATH 1444; (2) a suggestion of his to translate an occurrence of "en passant" with "obiter" rather than by "per transennam" is not matched by ATH 1444 (containing "per transennam" too);[157] (3) in ATH 1444 the Latin equivalent of 'mailles' addressed by De Raey is just omitted, and replaced by a series of dots;[158] (4) three phrases meaning, in Schuyl's and Clerselier's versions, the openings or intervals between the filaments (namely the only cases of the use of 'intervalle'/'intervallum' in an anatomical context in the treatise), and rendered in their editions with the use of the particle 'inter'/'entre', are rendered in ATH 1444

---

ce que l'humeur cristalline est un peu plus voûtée, et les autres"; ATH 1444, 91–92: "50.ª figura, quae parum est magis accurata, et inter alias partes oculi."

[151] Namely, there is a page missing before page 49 in the edition provided in this book.

[152] Van Roijen et al. 2013; Mourits 2016; Van Anrooij and Hoftijzer 2017. For a description and digitized version of the manuscript, see the digital collections of the Leiden Universiteitsbibliotheek: URL = <http://hdl.handle.net/1887.1/item:3487408>. Accessed 23 March 2024: "6 loose quires, 48 folios, 212 × 165 mm." No watermarks are observable in the digitized copy, consulted during the preparation of the present book.

[153] *Catalogus* 1657; *Catalogus* 1666; *Catalogus* 1668; *Catalogus* 1677; *Catalogus* 1739; *Catalogus* 1852; *Catalogus* 1879; some of the consulted copies (see the Bibliography) contain handwritten additions. Alas, I could not consult the list of the items held by the library compiled around 1778 by the curator Frederick Bernhard Albinus (1715–1778). The manuscript is now catalogued in Van Roijen et al. 2013, 77 (see also 9 and 101), based on the inventory of the library compiled by R. van Roijen in 1941 (where the manuscript has the catalogue number 293.05).

[154] For a reconstruction, see Van Roijen et al. 2013, 5.

[155] The reader can find samples of their textual reliance on *L'homme*, respectively, in Strazzoni 2023a, in Table A.1 in the present book, and in Monchamp 1886, 329–331 (see *supra*, n. 74). Their Latin texts cannot be traced to ATH 1444.

[156] ATH 1444, 19–20 and 30; signalled in the present edition.

[157] Cf. Descartes 1662, 13; Descartes 1664b, 10; ATH 1444, 12; signalled in the present edition. See *supra*, n. 71.

[158] Cf. Descartes 1662, 73; Descartes 1664b, 63; ATH 1444, 71; signalled in the present edition.

with phrases meaning openings within the filaments (viz. with the particles 'intra' and 'in').[159] So that ATH 1444 supports the idea that filaments are hollow even more than Schuyl's edition. Accordingly, considering De Raey's appreciation for Huyberts, we can also exclude that the latter was the translator behind ATH 1444.

In fact, it is notable that the equivalent of 'mailles' is not the only word replaced by dots in ATH 1444: in another case, a part of the word 'vendemiarum' is omitted in the same way ("....demiarum"), while in a third instance a word meaning food is omitted.[160] The case of 'mailles' can be explained in philosophical terms, as a difficulty in understanding the structure of the brain and filaments by the translator, who might have just left the matter unsettled, though also a copyist might have been puzzled by the text, and omitted such a word even if he found it in the manuscript he was transcribing. The case of "....demiarum," in turn, can be more clearly attributed to a copyist's difficulty in deciphering an already translated text (so that we can infer that the translator and the copyist of ATH 1444 were two different people, as mentioned above), while the case of the word standing for 'food' might have derived either from a difficulty in translating, or—more probably—in deciphering a text. Moreover, we need to take into account, as a complicating factor, that the omitted words themselves might have showed variants in the different copies to which the translator and the copyist of ATH 1444 had access. The matter is thus hard to settle, also because there is evidence that even the copyist of ATH 1444 had access to an at least partial French text, though this French text was not used to fill the spaces of the dots. This is revealed by at least two marginalia, both showing variants between the three versions and underlined in ATH 1444. In particular, we do find a Latin clarification based on a French version, as on the characterization of the external layer of the nerves as a membrane, skin or tunic (where the marginal correction matches Clerselier's version);[161] moreover, there is a French clarification in the description of the flowing of spirits out of the pineal gland: according to ATH 1444, they do so with a sort of down-up movement ("sourdent," "saliunt," also expressed in Schuyl's version with the use of the verb 'prosilio', or to jump—matching indeed the moving of spirits out of a spring, to which the gland is compared). On the other hand, in Clerselier's edition the movement of the spirits out of the gland is described by 'couler', meaning more the idea of flowing as such.[162] Moreover, a French phrase is provided (and underlined) in the main text itself, it being however unclear if it was employed by the copyist instead of an illegible text, or just left untranslated; notably, however, it shows a variant too, with respect to Schuyl's edition, so that it might have been the result of a slight re-working, which could have made difficult

---

[159] Cf. Descartes 1662, 72, 74, and 83; Descartes 1664b, 62, 64, and 74; ATH 1444, 71, 72 and 79; signalled in the present edition.

[160] Cf. Descartes 1662, 4 and 55; Descartes 1664b, 4 and 47; ATH 1444, 4 and 56; signalled in the present edition.

[161] Cf. Descartes 1662, 19: "membrana, sive tunica"; Descartes 1664b, 15: "la peau"; ATH 1444, 19: "tunica (…) [*in margine:* gallice pellis]." Underlining (here and henceforth) by the writer of the manuscript.

[162] Descartes 1662, 16: "[spiritus] prosiliant"; Descartes 1664b, 12: "[les esprits] coulent"; ATH 1444, 15: "[spiritus] saliunt [*in margine:* sourdent]."

its transmission across the manuscripts.¹⁶³ Further Latin changes and underlining (signalled in the present edition), in turn, reveal access to a text (or more) matching alternatively Clerselier's and Schuyl's versions: though, it is unclear whether a French manuscript, or even their editions, had been used to insert them.

Also—as evident from the edition of ATH 1444, and as I discuss further below—in a number of minor variants ATH 1444 adheres to Clerselier's rather than to Schuyl's version. At once, in the cases of more substantial variants it definitely adheres more to Schuyl's (still, with some exceptions).¹⁶⁴ Accordingly, we can reasonably interpret many minor variants between Schuyl's and Clerselier's editions as translation variants, while we do not have evidence whether we should attribute more substantial variants—for the most part, additional sentences to be found in Clerselier's version (again, with exceptions)¹⁶⁵—to Descartes's reworking of the treatise or to its editing by Clerselier.

One of the more relevant variants, for instance, concern the organs of smell. In *Principia* IV.193 Descartes identified olfactory organs in "two nerves or appendages of the brain (which do not protrude beyond the skull)," so that particles can be smelled if they "penetrate to these nerves through the pores of the spongy bone."¹⁶⁶ In turn, in his *L'homme* Descartes had identified the organs of smell in multiple (evidently, more than two) filaments originating from the basis of the brain and spreading towards the nose—but still not exiting from the skull. According to Clerselier's version only, such filaments are located below the so-called *processus mammillares*, or olfactory bulbs, which, prima facie, seem to be in fact the two appendages mentioned in the *Principia*.¹⁶⁷ Though both accounts might have been consistent with each other, as, notably, in his *Liber de osse cribriformi, et sensu ac organo odoratus, et morbis ad utrumque spectantibus* (1655) Konrad Victor Schneider (1614–1680) claimed— quoting from *Principia* IV.193—that for Descartes olfactory organs are not the two *processus mammillares*, but rather—as for Andrea Vesalio (1514–1564) and Jean Fernel (1497–1558)—some further *processus* or structures similar to nerves (which

---

¹⁶³ Descartes 1662, 17–18: "Dianam, quae forte se lavat"; Descartes 1664b, 13: "Diane qui se baigne"; ATH 1444, 17: "Dianam qui se baigne."

¹⁶⁴ Cf. Descartes 1662, 39: "(…) et quae poris destituuntur, illae ipsae, sicuti efficere non poterunt, ut anima ullum sentiat saporem (…)"; Descartes 1664b, 33: "(…) et qui n'ont point aussi de pores en elles-mêmes, où les petites parties de la langue, ou bien pour le moins celles de la salive dont elle est humectée, puissent entrer; comme elles ne pourront faire sentir à l'âme aucun goût (…)"; ATH 1444, 40: "(…) quique etiam ipsi nullos poros habent, per quos linguae particulae, vel ad minimum salivae, quibus humectatur ingredi possunt, quemadmodum enim non poterunt efficere ut anima ullum sentiat saporem (…)."

¹⁶⁵ Cf. Descartes 1662, 45–46: "crystallinus humor, quatenus visus noster in res propinquas, aut longe dissitas fertur, statim convexior redditur, aut planior, adeoque totam oculi figuram nonnihil immutat"; Descartes 1664b, 39–40: "sa figure se peut changer, et se rendre un peu plus plate, ou plus voûtée, selon qu'il est de besoin"; ATH 1444, 48: "hic humor pro intentione, qua visus noster in res propinquas aut longe dissitas fertur, mox in maiorem gibbum curvatus, mox [[p]] magis in planum porrectus, totam oculi figuram nonnihil immutat."

¹⁶⁶ Descartes 1982, 279; AT VIII/1, 318–319.

¹⁶⁷ Cf. Descartes 1662, 39–40; Descartes 1664b, 33–34; ATH 1444, 41–42; signalled in the present edition.

he also labels as mutilated nerves) located under such *processus*, and not exiting from the skull (whereas authors such as Galen, Avicenna and Ludovico Settala, 1552–1633, endorsed the idea that olfactory organs are the two *processus mammillares* themselves). A theory rejected by Schneider, for whom olfactory organs are multiple i.e. more than two filaments—comparable to nerves—entering the nose itself through the spongy bone, and thus exiting from the skull, as the nerves responsible for hearing and seeing do.[168] In turn, in a commentary on the *Principia* dating to 1659–1661 De Raey offered a partially different reading on IV.193, criticizing what he interprets as Descartes's identification of sense organs with the very *processus mammillares*, for the reason that these are located within the skull, whereas some *recentiores* claimed that the organs of smell are very subtle nerves reaching the nose.[169] In a later commentary (1664–1668), eventually, De Raey overtly endorsed Schneider's idea of quasi-nerves entering the nose, criticizing on this ground Descartes's identification of sense organs with the *processus mammillares* as given in *Principia* IV.193, but overtly claiming that Descartes himself acknowledged the existence of nerves exiting the skull and penetrating the nose in his *L'homme* (in whose extant versions, nonetheless, it is claimed that the mentioned filaments do not exit the skull).[170] To sum up, Clerselier might have used Schneider's interpretation of Descartes's text, differentiating the olfactory nerves or filaments from the *processus mammillares* (though without claiming that such filaments exit from the skull), as De Raey did by

---

[168] Schneider 1655, 8–9 and 107–109.

[169] "'Cerebri appendices': processus mammillares vocantur sed quia extra calvariam non protenduntur hae cerebri partes, explicatur difficile est quomodo odores eo perveniant. Ideoque recentiores quidam putant subtilissimos nervos alios per intimam narium superficiem dispersos inservire olfactui. Quae lis quia composita non dum est inter anatomicos, praestat hic ἐπέχειν i.e. assensionem cohibere," De Raey 1659–1661, 224. In his *Oratio inauguralis de gradibus et vitiis notitiae vulgaris circa contemplatione naturae et officio philosophi circa eandem* (1651), later re-issued in his *Clavis*, De Raey basically endorsed the view proposed by Descartes in his *Principia* that the olfactory organs are two appendices, not enveloped by any membrane, extended from the brain towards the nostrils (but without any specification whether they are extended outside the skull): "versus nares ex cerebro porriguntur duo nervosi processus, qui nudi incedunt," De Raey 1654, 10. At that point he most probably had not read *L'homme*.

[170] "'Cerebri appendices': anatomici processus mamillares vocant, sed non sine ratione dubitatur, an non subtiliores nervi qui comitantur hos processus mamillares penetrent in nares extra calvariam proprium organum huius sensus sint, quae sententia Keidneri [sic] est recentioris medici, author hos nervos in libro De homine etiam agnovit," Descartes 1664–1668, 89r. Notably, De Raey overtly mentions Descartes's *L'homme* only after its publication by Schuyl; as it is not mentioned in his 1659–1661 commentary (see *supra*, n. 169), and as one of his commentaries on Sennert can trace to 1661, we might hypothesize that he had access to the treatise only in 1661–1662; though there is evidence that he relied on it also in 1658, or at least that he had access to Huyberts's copy: see Sects. 1.4.2 and 1.4.3. So that his mention could have just been motivated by the circulation of the treatise in printed form; the same applies to his mentioning of Descartes's *Le monde*: mentioned in his 1658 commentary (in a section on the *Discours*) as a draft of the ideas expounded in his *Principia*, and whose publication (1664) is criticized, for this reason, in the commentary given on *Principia* III.47 in his 1664–1668 text: cf. De Raey 1658, 96: "(…) De lumine, ex quo haec Principia desumpta sunt. Author noster duplici usus est methodo: primo rudes conceptus delineavit, secundo, eos in ordinem redegit"; De Raey 1664–1669, 45r: "[v]ocatur tractatus ille (Gallice conscriptus) De lumine, ab imprudentibus editus, cum primae et rudes tantum cogitationes sint, ex quibus hoc opus est."

interpreting Descartes's text as meaning that olfactory nerves are, as for Schneider, those present in the nose. Still, the variant present in Clerselier's edition can be just attributed to Descartes himself, who could have relied on Vesalio and Fernel.

Another interesting group of major variants concern Descartes's explanation of muscular movement: the specific process, as mentioned in Sect. 1.3, the account of which was plagiarized by Regius. The explanation is decidedly more developed in Clerselier's version,[171] even if without a figure it is still quite hard to understand, especially in the functioning of the valves that regulate the flow of spirits in the nerves and muscles (and Regius indeed provided a simplified version of the process).[172] Clerselier himself remarked, in his 1664 preface, how Descartes's text "says many things in few words," and tried to clarify his model by a discussion of the autograph figure of the nerve-muscle system in his possession and of those prepared by Van Gutschoven and La Forge.[173] An intervention of his in clarifying the text itself is thus probable, though precisely the scandal caused by Regius might have led Descartes to intervene on the text precisely in such an explanation. Notably, other variants concern Descartes's theory of the passions, namely another topic for the treatment of which Regius probably relied on *L'homme*, and which Descartes discussed with Elisabeth, re-directing her to his treatise in his letter of 6 October 1645 (as seen in Sect. 1.3).[174] Similarly, important variants concern the pineal gland,[175] an idea which Descartes came to develop especially in his *Les passions de l'âme* (1649), viz. that came to be considered in more detail, along with the treatment of the passions, in his last years. In conclusion, we can reasonably assume that Descartes came to rework his text after its copying in *VanSurck* (a pre-revision text resulting also in ATH 1444).

As mentioned above, in the case of minor variants ATH 1444 agrees more than once with Clerselier's rather than with Schuyl's version: provided that ATH 1444 originated from *VanSurck*, we can thus attribute a consistent part—though not all— of such variants (in Schuyl's editions, 1662 and 1664) to Schuyl's translating and editing, and to his mistakes (especially in the first edition) or those of the copyists of the manuscripts he used (as, for instance, in the case of the mistaking of 'deux' and 'd'eux').[176] Indeed, the two editions of Schuyl contain variants with respect to each other,[177] with the text of the second edition coming to match (after corrections and

---

[171] Cf. Descartes 1662, 21–24 and 26–29; Descartes 1664b, 17–21 and 23–25; ATH 1444, 21–24 and 27–28; signalled in the present edition. ATH 1444 reveals, by some underlining, changes, and additions, the centrality of the issue.

[172] Strazzoni 2023a.

[173] Descartes 1664b, *Préface*, 14–24 (unnumbered).

[174] Descartes 1662, 69–70; Descartes 1664b, 59; ATH 1444, 67–68; signalled in the present edition.

[175] Cf. Descartes 1662, 75 and 101; Descartes 1664b, 66 and 89; ATH 1444, 73 and 94; signalled in the present edition. Cf. also Descartes 1662, 105: "(…) et proinde spiritus adducerent (…)"; Descartes 1664b, 93: "(…) et par conséquent qu'ils pussent recevoir des esprits d'autres points de la glande que de ceux qui sont marqués a, b, c, et les conduire (…)"; absent in ATH 1444.

[176] Cf. Descartes 1662, 76; Descartes 1664b, 67; ATH 1444, 74; signalled in the present edition.

[177] Considered in Meschini 2011, n. 27. Also the two editions by Clerselier (Descartes 1664 and Descartes 1677b) contain some variants (signalled in the present edition), which can be interpreted

additions) those of ATH 1444 and Clerselier:[178] again, with exceptions, as in some cases the corrections match only Clerselier's version,[179] or are different from all the other versions.[180] Such corrections and additions were made, most probably and for the most part, on the basis of the manuscripts Schuyl had at his disposal at the time of his first edition and afterwards (*VanSurck*, *Schuyl*, and *Heereboord*), as there is no evidence that he had access to Clerselier's edition (which probably appeared shortly after Schuyl's second edition),[181] or to the text of ATH 1444. Indeed, even if the translations of Schuyl and that of ATH 1444 reveal textual agreements (along with variants present in the same period),[182] these can in principle be explained as due to the fact that we are dealing with translations of, generally speaking, the same text. Namely, at the present state of research we cannot deem the translation of ATH 1444 to be based on that of Schuyl (or vice-versa), even it is reasonable that it was prepared before its publication, when a Latin version of the treatise was not publicly available. However, we cannot exclude (as mentioned above), that the copyist of ATH 1444 had access to Schuyl's or Clerselier's version (as in the case of 'son' and 'ton' discussed in Sect. 1.5, a variant which might also have been an editorial or translation one), or to further manuscripts matching them.

As far as the variants unique to ATH 1444 are concerned, we can interpret them for the most part as translation variants, among which we can count its beginning with "Homines." According to Clerselier, indeed, the beginning of the treatise was subject to a translation variant in Schuyl's edition, starting with "Omnes homines": so that most probably Clerselier could read in the copy *Clerselier1* "Ces hommes," viz. the same beginning of his edition, and claimed that Schuyl changed it. Accordingly, also the translator of ATH 1444, puzzled by the demonstrative 'ces', might have decided to start with "Homines": probably ignoring, like Schuyl (as reported by Clerselier), that the autograph bore the heading "Chapitre 18," being a continuation of Descartes's *Le monde*. The latter text, indeed, consists of 15 chapters, with the missing chapters 16 and 17 apparently devoted to the treatment of plants and animals, as declared in the

---

as corrections based on the same autograph, corrections of typos, or clarification editorial in nature, viz. not based on other versions. See Van Otegem 2002, volume 2, 524.

[178] Cf. Descartes 1662, 90, 92, 95, 97, and 101; Descartes 1664a, 90, 92, 95, 97, and 101; Descartes 1664b, 80, 84, 86, and 88; ATH 1444, 84, 85, 88, 91, and 93; signalled in the present edition.

[179] Cf. Descartes 1662, 93, 95, 97, and 104; Descartes 1664a, 93, 95, 97, and 104; Descartes 1664b, 82, 84, 86, and 92; ATH 1444, 86, 88, 91, and 96; signalled in the present edition.

[180] Cf. Descartes 1662, 92; Descartes 1664a, 92; Descartes 1664b, 80; ATH 1444, 85; signalled in the present edition.

[181] Schuyl's *Epistola* opening his 1664 edition is dated 6 March 1664, and reports that the book was already in press and could not be corrected (see *supra*, n. 81), while Clerselier's edition was finished being printed on 12 April 1664: Descartes 1664b, *Préface*, 62 (unnumbered).

[182] Cf. Descartes 1662, 46: "[d]enique O O sunt sex aut septem musculi extrinsecus oculo affixi, quibus facillime quaquavorsum moveatur"; Descartes 1664b, 40: "[e]nfin O, O, sont six ou sept muscles attachés à l'œil par-dehors, et qui le peuvent mouvoir très facilement et très promptement de tous côtés"; ATH 1444, 34: "[d]enique O O sunt 6 aut septem musculi extrinsecus oculo affixi, quorum ope promptissime et facillime quaquaversum moveri potest."

1.8 Conclusion   45

Discours, where such topics are described as having already been dealt with.[183] Other variants unique to ATH 1444 (sometimes variants within variants, if we compare it to the versions of Schuyl and Clerselier),[184] in turn, might have been interpolations more editorial in nature, like—as mentioned above—a reference to the condition of one of the figures (as *VanSurck*, from which it certainly derived, was figureless), or as the additions which cannot be explained as translation variants, and revealing a possible access to further manuscripts. Further, frequent variants are due to the mistaking of words by the translator (as 'donc'/'dont', or 'ces'/'ses', 'les'/'ses'), to apparent errors in the translation of sentences and/or omissions hindering their full comprehension,[185] or to transcription errors (as signalled in the appended edition, where, moreover, I have suggested a reading of certain words).

Last but not least, it is worth noting that in the cases of evident ambiguities, deletions, corrections, additions or omissions (by dots) we do find the use of marginal symbols and underlining, most probably to be traced to the copyist of ATH 1444 itself, who might have reflected on its state, and confronted it with a French text, as seen above.

1.8 Conclusion

As a conclusion, it is worth summarizing the (hypothetical) paths of the manuscript dissemination of *L'homme* in its main lines and actors. Around 1641–1642, Descartes allowed a first, figureless version of the treatise, conveyed by a messy manuscript, to be transcribed by Van Surck: from whose copy Pollot (1643–1646) and Elisabeth of Bohemia (1644–1645) made, or had further copies made, unchecked by Descartes and with a prohibition of further dissemination. This notwithstanding, it was Elisabeth who probably allowed the further dissemination of the treatise, through Johnson and Regius: the latter being in possession of or having access to a copy in 1646. Later, the treatise had a broad dissemination in the Dutch areas, as it was owned by Heereboord (with no evidence on when this happened), De Raey (who at least had access to it, from 1652 onwards, and who at some point prepared a Latin translation of it), and Huyberts (who relied on its contents in 1652, and worked on the preparation of its figures, if not textual editing, in 1656). We can exclude that De Raey's translation has survived as the text of ATH 1444, and there is no evidence to relate the efforts of

---

[183] Cf. Descartes 1662, 1; Descartes 1664b, *Préface*, 3 (unnumbered) and 1; ATH 1444, 1; AT VI, 45.

[184] Cf., for instance, Descartes 1662, 74: "[u]t vero omnia explicem, quae in illo plexu occurrunt notatu digna"; Descartes 1664b, 66: "mais afin que je puisse plus commodément expliquer toutes les particularités de ce tissu"; ATH 1444, 72: "[u]t autem commodius et facilius particularitates omnes istius texti explicare possim, (…)."

[185] Cf., for instance, Descartes 1662, 40: "crassioribus iis, quas superius odores nominavi"; Descartes 1664b, 34: "qui soient plus grosses que celles que j'ai ci-dessus nommées odeurs"; ATH 1444, 42: "quae sint crassioribus, quas propterea odores."

Huyberts to such a text, which was in any case probably prepared before the appearance of Schuyl's first edition (1662). In fact, De Raey's appreciation for Huyberts's work suggests that the latter might not have been its translator either (as the text of ATH 1444 was not immune to the criticisms moved by De Raey to Schuyl's edition); moreover, as ATH 1444 contains a remark on the quality of one of the figures, we can suppose that its translator and illustrator were two different persons. Around 1656, moreover, the Elzeviers were attempting to organize its publication, including the figures (if not the editing) by Huyberts, after having asked Chanut about the treatise around 1651–1653. Chanut, in fact, had to be the first editor of Descartes's posthumous writings, before Clerselier undertook this task in 1654. As late as in July 1654, anyway, Clerselier was not in possession of any manuscript, while he might have received one copy from Andreae before or during December of the same year. Chanut, in turn, most probably obtained the autograph Clerselier later used for his edition—an autograph including changes made by Descartes on the first version of his treatise—between mid-1653 and late 1655, delivering it to Clerselier maybe after having also provided him with a copy deriving from that of Elisabeth, which he got in 1653 with the help of Chevreau (who got a copy through La Voyette). Also, Clerselier received a copy from Schuyl (probably in 1660–1661), and, with his autograph, prompted a further dissemination of the treatise: to Huyberts again (in 1657), Van Gutschoven (1660), La Forge (1660), and, probably through the former, Philippi (1660–1661). Eventually, for his editions Schuyl used a copy of the copy of Pollot, and the copy of Van Surck; moreover, he also had at his disposal the copy of Heereboord.

Chapter 2
Edition of ATH 1444

Abstract In this chapter I provide an edition, with an apparatus of different readings and comments, of the partial text of René Descartes's *Traité de l'homme* conveyed in a Latin translation by the manuscript ATH 1444, now extant at the Bibliotheca Thysiana at Leiden (part of the Leiden Universiteitsbibliotheek), and titled *Tractatus de homine a Cartesio*. The readings are considered with respect to the Latin and French editions published by Florentius Schuyl (1662 and 1664) and Claude Clerselier (1664 and 1677). This transcription offers an overview of the differences between the foremost known versions of the treatise, tracing to (possible) author's and editors' interventions, with particular attention to those concerning conceptually loaded terms, errors in translation or transcription, references to (now) missing figures, and stylistic choices.

Keywords René Descartes · *Traité de l'homme* · *Tractatus de homine a Cartesio* · Ms. ATH 1444 · Early modern handwritten sources · Edition with different readings

In what follows I offer an edition of ATH 1444, provided with (1) an apparatus of different readings, which I will hereafter refer to as 'variants', and textual comparisons taking into account the editions of Descartes's *L'homme* by Schuyl (1662 and 1664) and Clerselier (1664 and 1677), and (2) an apparatus of notes commenting upon the variants themselves and the peculiarities of the text. As the comparison is between two different Latin translations and a French version, I do not offer a critical edition aimed at the reconstruction of an archetype, but rather a textual comparison aimed at revealing the different ways in which Descartes's *L'homme* had been translated and edited, and which can serve to highlight how it could have been re-worked by Descartes during his lifetime (as discussed in Chap. 1).

As to the variants and the comparisons between the texts—which are presented by footnotes numbered with Arabic numerals, while endnotes, numbered with Roman numerals, contain comments on certain words or phrases—I have followed these criteria:

1. I have reported major variants, viz. those revealing that certain parts of the text were added or missing in its different versions;
2. I have reported the variants between the two editions of Schuyl and Clerselier: these are indicated, respectively, with 'S1' (Descartes 1662) and 'S2' (Descartes 1664a), 'C1' (Descartes 1664b) and 'C2' (Descartes 1677b); the page numbering of S2 follows that of S1, so that only the page numbers of S1 have been provided;
3. I have reported minor variants (especially apparent translation variants) which

    - ease or complicate the comprehension of the text, as in the case of the use of pronouns instead of nouns, or of different conjunctions between periods;
    - reveal different translations of the same term in a given text;
    - concern philosophical and scientifically-loaded terms;
    - reveal possible access of Schuyl to ATH 1444 or vice-versa;
    - concern references to (missing) figures, e.g. differences in the use of capitalization for reference letters, or specification that letters or numbers are being used;
    - reveal that ATH 1444 was following either Schuyl's or Clerselier's version of the text;
    - reveal or shed light on ambiguities and possible errors in translation or transcription;
    - are samples of translation variants per se, viz. they are stylistic in nature;

4. I have compared the texts (regardless of their containing variants) concerning

    - not clearly readable text;
    - underlined text;
    - deletions, omissions, additions, corrections revealing a potential access to further manuscripts or texts, or philosophically relevant;

5. the use of ellipses in comparisons means that the omitted text does not contain variants relevant to the comparisons in question; in some cases, the compared texts have been interpolated to facilitate understanding;
6. the position of marginal, subscript, or superscript text relevant for the variants has been sometimes changed in the apparatus, for clarity;
7. I have omitted to provide all those parts of text clearly provided by Schuyl and Clerselier in their editions, as marginal indications of figures, titles of paragraphs or parts; however, the paragraph numbers have been indicated.

Conventions adopted in the edition:

1. the deleted text has been put between brackets [[]];
2. the text and symbols added in the margins or as subscript and superscript have been put between brackets [], and indicated as such; I have indicated, in the notes, to which part of the main text they (might) refer;
3. dubious text is put between brackets { }: I have provided the deleted or dubious text, resolving if possible the abbreviations and contractions; otherwise, I have used an ellipsis ... instead of each illegible word or part of word;
4. my interpolations are put between brackets [];
5. reference symbols (∧, //, #) have been modernized;
6. the copyist's use of series of dots (for omissions) or lines (for unifying the text) has been rendered as a series of dots or a line matching the length of the ones present in the manuscript;
7. foot-page words used to connect the text of different pages have been omitted;
8. underlining and indenting have been kept as in the original;
9. punctuation, capitalization and diacritical signs have been only slightly modernized;
10. abbreviations and contractions (including deleted ones) have generally been resolved without any indication of change; the most common have been left unresolved, and dubious ones have been interpolated;
11. correct or alternative readings of obvious misspellings and mistakes have been suggested in the notes or by interpolation;
12. series of letters and numbers mentioned in the text and referring to (missing) figures have been kept unseparated, unless they were clearly separated in the manuscript by spaces or dedicated punctuation, or when the text clearly refers, through them, to separate entities; full stops usually accompanying them have been omitted.

[1, unnumbered]

Tractatus
De Homine.
a Cartesio[1],[i]

[2]Homines[3] componuntur ex anima et corpore quemadmodum nos: oportet igitur primum corpus sigillatim, ⟦ad⟧ ac deinde animam[4] describam, et tandem ostendam quomodo duae istae naturae coniungi atque uniri debeant, ut constituant homines nobis similes.

[5]Suppono corpus nihil aliud esse quam statuam et machinam terrestrem quam Deus expresse sic format ut eam [*subscriptum:* ∧] quantum [*in margine:* ∧ nobis] fieri potest similem reddat,[6] adeo ut non tantum extrinsecus illam exornet[7] colore et figura omnium nostrorum membrorum, verum etiam intrinsecus recondat omnes partes requisitas ut eat, edat,[8] respiret, ac tandem imitetur omnes nostras functiones quae possunt fingi procedere a materia, et a sola organorum dispositione dependere. [9]Videmus {enim}[10] horologia fontes arteficiales, molas aliasque his similes machinas quae licet tantum [2] ab hominibus factae vim tamen habent sponte sese diversimode moveant.[11] Et existimo nullatenus tam varios me posse imaginari motus in his quam quidem suppono a Deo factos[12] nec illis tribuere tot arteficia ut occasionem non habeas cogitandi adhuc plura esse posse. [13]Non itaque hic diutius insistam in describendis ossibus, nervis, musculis, venis, arteriis, stomacho, hepate, corde,[14] cerebro, aut aliis diversis partibus quibus eam componi oportet: illas enim suppono omnimodo similes partibus nostrorum corporum, quae et eadem sortitae

---

[1] Tractatus De Homine. a Cartesio] Renatus Des Cartes De homine S1, 1; L'Homme de René Descartes C1, 1.

[2] § 1 S1; § 1 and beginning of part 1, C1, 1.

[3] Homines] Omnes homines S1; Ces hommes C1.

[4] animam] separatim animam quoque S1; l'âme aussi à part C1.

[5] § 2 S1; § 2 C1.

[6] ut eam [*subscriptum:* ∧] quantum [*in margine:* ∧ nobis] fieri potest similem reddat,] nostro corpori persimilem; S1; pour la rendre la plus semblable à nous qu'il est possible: C1.

[7] exornet] exornantur, sit conspicua S1; donne C1.

[8] eat, edat,] ut edat, bibat S1, 2; qu'elle marche, qu'elle mange, C1, 2.

[9] § C1.

[10] Videmus {enim}] Certe quia S1; Nous voyons C1.

[11] sponte sese diversimode moveant.] sese velut sua sponte movendi S1; de se mouvoir d'elles-mêmes en plusieurs diverses façons; C1.

[12] in his [machinis] quam quidem suppono a Deo factos [motus]] quibus [motibus] illa machina, quam a Deo factam suppono, agitetur S1; en celle-ci [machine], que je suppose être faite des mains de Dieu C1.

[13] § C1.

[14] corde,] cordis, S1; la rate, le cœur, C1.

fuere nomina, et quas si non sufficienter cognoveris anatomicum consulere licet;[15] et quod ad eas quae propter suam exiguitatem visum fugiunt, clarius[16] demonstravero motibus, qui inde dependent: ita ut hic tantum necesse sit ordine explicem motus illos, ac eadem opera dicam quaenam sint nostrae functiones[17] quas representant.

[18]Primo[19,ii] alimentum concoquitur in stomacho huius machinae vi quorundam liquorum, qui se in illius partes insinuantes easdem separant, [3] ⟦agi⟧ agitant, et[20] calefaciunt, quem⟦{...}⟧admodum [*in margine:* #][iii] communis aqua partes calcis,[21] aut aqua fortis partes metallorum,[22] quin et[iv] liquores hi prompte admodum a corde per arterias deducti (u⟦p⟧t postea dicam) non possunt quin sint ⟦calidi⟧ valde calidi, imo cibus, eius ut plurimum esse natura, ut sponte[23] corrumpi et calefieri possit, ut videre est in foeno recenti quod antequam siccum fuerit in horrea conditur, [24]et scito agitationem, quam recipiunt particulae illae[25] cibi, iunctam agitationi stomachi et intestinorum, quae partes illas continent, et dispositioni parvorum filamentorum, ex quibus intestina componuntur, efficere ut simulac concoquuntur descendant paulatim ad illum ductum, per quem exire debent crassiores,[26] ac interim subtiliores ⟦{a}⟧et magis agitatae offendant circumquaque infinitos exiguos poros, per quos fluunt in ramusculos magnae venae easdem deducentis versus iecur,[27] cum tamen nihil sit propter exiguitatem pororum, quod eas[28] a crassioribus separet:[29,v] quemadmodum ⟦quemadmodum⟧ farinam agitando in sacco, [4] quod subtilius enim effluit,[30] ac

---

[15] anatomicum consulere licet;] anatomico discere licet: saltem illas, quae sua magnitudine visui manifestae sunt. S1; pouvez vous faire montrer par quelque savant anatomiste, au moins celles qui sont assez grosses pour être vues, si vous ne les connaissez déjà assez suffisamment de vous-même: C1.

[16] clarius] haud obscure S1; plus facilement et plus clairement C1.

[17] quaenam sint nostrae functiones] functiones S1; quelles sont celles de nos fonctions C1.

[18] § 3 S1, 3; § 3 C1.

[19] Primo] igitur S1; Premièrement C1.

[20] et] et tandem S1; et C1, 3.

[21] partes calcis,] calcem vivam, S1; celles de la chaux vive, C1.

[22] partes metallorum,] metalla S1; celles des métaux. C1.

[23] sponte] sponte citra ullam alterius rei mixturam S1; toutes seules, C1.

[24] § C1.

[25] illae] incalescentes illae S1; ces viandes en s'échauffant C1.

[26] per quem exire debent crassiores,] quem natura crassioribus cibi partibus excernendis destinavit, S1; par où les plus grossières d'entre elles doivent sortir; C1.

[27] iecur,] hepar. S1; le foie, et en d'autres qui les portent ailleurs, C1.

[28] eas] tenuiores cibi particulas S1; les C1.

[29] separet:] segregaverat: S1; separent [*errata corrige:* separe] C1.

[30] subtilius enim effluit,] purior tenuiorque eius pars per saccum excutitur, relictis furfuribus; S1, 4; toute la plus pure s'écoule, C1.

nihil est quam pororum exiguitas per quos fluit, quod impediat quo minus reliquum subsequatur.

³¹Subtiliores istae cibi particulae inaequales cum sint, et valde imperfecte mixtae liquorem componunt, qui prorsus turbidus, et subalbidus maneret, nisi misceretur³² statim cum sanguinis massa contenta in ramusculis venae portae, quae eum ex intestinis in os ramusculos venae cavae deducit, [[{...}]]et haec rursus [*in margine:* NB] versus cor et [*superscriptum:* in] iecur³³,ⱽⁱ quasi in unicum vas:³⁴ ³⁵quin etiam hic notandum poros iecinoris sic esse constitutos, ut cum liquor ille, ipsum intrat ibi subtilisetur, elaboretur, coloretur,³⁶ ac formam sanguinis adipiscatur. Quemadmodum succus atrarum uvarum, qui albus est, in rubellum vinum convertitur, cum [*in margine:* //]ᵛⁱⁱ ....demiarumᵛⁱⁱⁱ tempore in cupa cum scapis fermentari sinunt.³⁷ ³⁸Caeterum sanguis sic contentus in venis unicum tantum exitum manifestum habet per quem egredi possit, nempe eum qui ipsum in dextram concavitatem [5] cordis ducit. Nota etiam³⁹ [[carnem cordis]]⁴⁰ [*in margine:* //]ⁱˣ in suis poris continere quendam ex illis ignibus absque lumine, de quibus egimus antea,⁴¹,ˣ qui idipsum tam calidum et ardens reddit, ut simulac intrat sanguis, in alterutam eius concavitatem⁴² illico intumescat et dilatetur ut poteris experiri in sanguine aut lacte alicuius animalis,⁴³ si guttatim infundatur in vas valde calidum. Ignis autem qui est in corde machinae, quam describo, nulli alii rei inservit, quam sic dilatando, calefaciendo, et subtilisando sanguini, qui continuo per tubulum venae cavae guttatim decidit in concavitatem dextram,⁴⁴ unde exhalat in pulmonem, et inde per venam, quam

---

³¹ § 4 C1.

³² misceretur] permisceretur S1; une partie se mêle C1, 4.

³³ contenta in ramusculis venae portae, quae eum ex intestinis in os ramusculos venae cavae deducit, [[{...}]]et haec rursus [*in margine:* NB] versus cor et [*superscriptum:* in] iecur] contento in ramis venae portae. Haec vero illum ex intestinis in hepar, adeoque in plurimos ramos venae, quam cavam vocant, deducit. Quae porro eundem in cor S1; contenue dans tous les rameaux de la veine nommée porte (qui reçoit cette liqueur des intestins) dans tous ceux de la veine nommée cave (qui la conduit vers le cœur) et dans le foie C1.

³⁴ vas:] vas transfert. S1; vaisseau. C1.

³⁵ § C1.

³⁶ coloretur,] colorem (...) adipiscatur. S1; prend sa couleur, C1.

³⁷ cum ....demiarum tempore in cupa cum scapis fermentari sinunt.] cum vindemiarum tempore scapis suis permixtus, vasi ad fermentandum committitur. S1; lorsqu'on le laisse cuver sur la râpe. C1.

³⁸ § 5 C1.

³⁹ Nota etiam] Scire quoque operae pretium est, S1; Et sachez C1.

⁴⁰ [[carnem cordis]]] parenchymatis cordis S1; la chair du cœur C1.

⁴¹ egimus antea,] alibi S1, 5; je vous ai parlé ci-dessus, C1.

⁴² concavitatem] ventriculum S1; chambres ou concavités qui sont en elle, C1.

⁴³ alicuius animalis,] cuiusvis animalis S1; quelque animal que ce puisse être, C1.

⁴⁴ concavitatem dextram,] dextrum cordis ventriculum S1; la concavité de son côté droit, C1.

anatomici vocant arteriam venosam, in alteram eius concavitatem, unde per totum corpus dispergitur.

⁴⁵Caro pulmonis adeo rara et mollis est, attamen sic a respirationis⁴⁶ aere frigefacta, ut simulac vapores sanguinis e concavitate dextra cordis egredientes, pulmones intrant per arteriam, quam anatomici vocant [6] venam arteriosam, ibi condensantur, et rursus in sanguinem avertantur, postea inde in sinistram cordis concavitatem guttatim decidant, quo si intrarent antequam rursus fuissent condensati non sufficerent nutritioni ignis, qui ibi est.

⁴⁷Et sic vides respirationem quae huic tantum machinae inservit, condensandis illis vaporibus, non minus [in margine: //]ˣⁱ necessariam esse nutritioni illius ignis, quam ⟦q⟧ eam quae in nobis et vitae nostrae conservationi, adminimumˣⁱⁱ nobis qui sumus formati homines,⁴⁸ nam quod ad infantes, qui dum adhuc sunt in utero matris respirando aerem recentem attrahere nequeunt, duos ductus habent qui hunc defectum supplent, unum per quem sanguis venae cavae transit in venam arteriosam,⁴⁹ et alterum, per quem vapores arteriae venosae⁵⁰ exhalant in magnam arteriam, ⁵¹et quod ad illa animalia, quae prorsus pulmone carent, unam tantum concavitatem⁵² ⟦hut⟧ˣⁱⁱⁱ habent aut si plures continuae⁵³ sunt.

[7] ⁵⁴Pulsus⁵⁵ arteriarum dependet ab undecim pelliculis quae instar valvularum claudunt et aperiunt introitus⟦,⟧ quatuor vasorum quae duas cordis concavitates respiciunt, nam uno pulso⁵⁶ cessante alter pulsare ⟦{...}⟧ paratus est, valvulae quae sunt in introitu duarum arteriarum exacte clausae sunt, et illae quae sunt in introitu duarum venarum apertae, ita ut non possint, quin statim cadant duae guttae sanguinis e venis in quamque cordis concavitatem.⁵⁷

ˣⁱᵛDeinde simul guttae illae sanguinis rarefiunt et extenduntur in spatium multo maius quam antea occupaverant, ac si pellunt et claudunt valvulas in introitu duarum venarum sitas, hoc modo impedientes ne plus sanguinis⁵⁸ succedere possit; et pellunt et aperiunt duarum arteriarum valvulas per quas prompte, et vi egrediuntur, et sic

---

⁴⁵ § 4 S1; § 6 C1, 5.

⁴⁶ respirationis] beneficio inspirationis S1; de la respiration C1.

⁴⁷ § C1.

⁴⁸ nobis qui sumus formati homines,] nos, qui membris consummatis constamus, S1, 7; en ceux de nous qui sont hommes formés: C1.

⁴⁹ venam arteriosam,] venam, quam arteriam vocant, S1; la veine nommée artère, C1.

⁵⁰ arteriae venosae] arteriae, quam venam appellant S1; l'artère nommée veine C1.

⁵¹ § 5 S1.

⁵² concavitatem] in corde ventriculum S1; concavité dans le cœur C1.

⁵³ continuae] continui S1; toutes consécutives l'une à l'autre. C1.

⁵⁴ § 6 S1; § 7 C1.

⁵⁵ Pulsus] Pulsus S1; Le pouls, ou battement C1.

⁵⁶ uno pulso] pulsuum horum uno S1; au moment qu'un de ces battements C1, 5–6.

⁵⁷ in quamque cordis concavitatem.] Altera videlicet in dextrum: altera in sinistrum cordis ventriculum. S1; une dans chaque concavité du cœur. C1.

⁵⁸ sanguinis] sanguinis copia cor S1, 9; sang dans le cœur C1, 6.

[[{...}]] efficiunt, ut cor et reliquae arteriae corporis[59] eodem momento intumescant, sed statim sanguis ille rarefactus iterum condensatur, aut alias partes [8] penetrat, quo fit ut cor et arteriae detumescantur, et valvulae in introitu arteriarum[60] sitae rursus [[cl]] claudantur, et in venarum introitu sitae,[61] rursus aperiantur, et transitum praebeant duabus aliis guttis sanguinis quibus cor et arteriae denuo intumescuntur, eodem modo ut praecedentibus.

[62]Cognita sic causa arteriarum pulsus,[63] facile est intellectu non esse sanguinem contentum in venis huius machinae qui immediate ab hepate provenit, quam qui contentus est in arteriis, et qui [[ia]] iam in corde distillatus fuit, qui aliis eius partibus agglutinari possit, et eis particulis reparandis[64] inservire, quas continua eius[ccxxxviii] agitatio, et diversae aliorum corporum ei circumstantium[ccxxxix] actiones depellunt.[65] Sanguis enim in venis contentus fluit paulatim[66] ab illarum extremitatibus versus cor secundum dispositionem quarundam valvularum,[67] quas anatomici animadverterunt [9] variis in locis iuxta venas, quo nobismet satis persuadere possumus[68] simile in nobis fieri, contra autem pellitur ille sanguis qui in/arteriis[xv] continetur vi quadam ac diversis concutionibus [*superscriptum:* conatibus][69,xvi] versus extremitates, ut sic facile possit iungi et uniri omnibus eius membris, eaque hoc modo alere, imo et augere, si machina[70] repraesentet corpus hominis ad id dispositum.

[71]Nam simulac arteriae intumescunt particulae sanguinis quas continent progrediendo huc et illuc offendunt radices quorundam filamentorum, quaeque egredientes ex extremitatibus ramusculorum arteriarum componunt ossa, carnes, cutem, nervos, cerebrum, et reliqua membra solida, secundum diversos modos quibus coniunguntur intertexuntur et sic vim habent ea aliquantulum propellendi[72] et se in illius loca insinuandi. Postea, simulac arteriae detumescunt quaeque particularum ubi sunt manent,[73] atque hac sola [10] causa iungitur et unitur iis quae attingunt, iuxta id

---

[59] reliquae arteriae corporis] totius corporis arteriae S1; toutes les artères du corps C1.

[60] arteriarum] duarum arteriarum S1; des deux artères C1.

[61] et in venarum introitu sitae,] aliae vero in duarum venarum introitu S1; et celles qui sont aux entrées des deux veines C1.

[62] § 7 S1; § 8 C1.

[63] arteriarum pulsus,] pulsus arteriarum S1; pouls, C1.

[64] eis particulis reparandis] restauretur quicquid S1, 10; à réparer ce C1.

[65] depellunt.] expellunt. S1; détachent et font sortir; C1, 7.

[66] paulatim] sensim S1; toujours peu à peu C1.

[67] valvularum,] valvularum S1; petites portes, ou valvules, C1.

[68] nobismet satis persuadere possumus] manifeste satis constet S1; vous doit assez persuader C1.

[69] diversis concutionibus [*superscriptum:* conatibus]] pluribus pulsibus S1; diverses petites secousses C1.

[70] machina] machina S1, 11; elle C1.

[71] § 8 S1; § 9 C1.

[72] ea aliquantulum propellendi] nutrienda illa membra aliquantulum distendendi S1; les pousser quelque peu devant soi C1.

[73] quaeque particularum ubi sunt manent,] singulae sanguinis particulae spatiis, quae occuparunt, insertae subsistunt. S1; chacune de ses parties s'arrête où elle se trouve, C1.

quod supra[74],[xvii] dictum, [75]si itaque sit corpus infantis quod machina⟦m⟧ nostra⟦m⟧ repraesentat, materia eius adeo erit mollis, et pori eius tam facile poterunt dilatari ut particulae sanguinis quae compositionem membrorum solidorum ingredientur, ut plurimum paulo crassiores futurae sint,[76] imo fiet ut duae tresve simul succedant uni,[77] quae erit causa eius accretionis, sed interim materia eius[78],[xviii] membrorum paulatim indurescet, sic ut post elapsos annos aliquot pori eius nequaquam poterint dilatari, et sic crescere desinens repraesentabit corpus hominis adulti.

[79]Caeterum paucae[80] sunt sanguinis partes quae unaquaque vice solidis membris uniri possunt iuxta modum quem modo descripsi, ast pleraeque[81] in venas rediunt per extremitates arteriarum, quae variis in locis iunctae sunt vena[11]rum extremitatibus,[82] indeque particulae aliquae fortasse nutritionem quorundam membrorum ingrediuntur, sed maior pars ad cor revertitur, postea inde iterum in arterias, ita ut motus sanguinis in corpore nihil aliud fit[xix] quam perpetua quaedam circulatio; [83]caeterum quaedam particulae sanguinis lienem ingrediuntur, quaedam vesiculam fellis, quam[84] immediate ex arteriis, quaedam in stomachum et intestina fluunt, ubi sunt loco aquae fortis ut facilitent concoctionem ciborum, et quia illuc deducuntur quasi in momento ex corde per arterias, non possunt quin sint valde calidae, qui fit ut earum vapores facile ascendere possint per guttur in os ibique componere salivam: quaedam[85] etiam renes[86] perfluendo in urinas mutantur aut in sudorem, aut alia excrementa per poros cutis.[87] Et ubique semper aut situs [*subscriptum:* ∧; *superscriptum:* ∧ aut] figura[88] aut exiguitas pororum[89] [12] efficit ut unae potius quam

---

[74] supra] alibi S1; ci-dessus. C1.

[75] § C1.

[76] crassiores futurae sint,] maiores sint illis, quarum loca occupabunt. S1; seront (…) plus grosses que celles en la place de qui elles se mettront, C1, 7–8.

[77] uni,] in locum unius: S1; à une seule, C1.

[78] eius] illorum S1; ses C1.

[79] § 10 C1, 8.

[80] paucae] non nisi paucae admodum S1, 12; il n'y a que fort peu C1.

[81] ast pleraeque] Pleraeque S1; mais la plupart C1.

[82] per extremitates arteriarum, quae variis in locis iunctae sunt venarum extremitatibus,] per varias arteriarum venarumque anastomoses. S1; par les extrémités des artères, qui se trouvent en plusieurs endroits jointes à celles des veines. C1.

[83] § 11 C1.

[84] vesiculam fellis, quam] fellis vesiculam. Et rursus tam ex liene, atque ex vesicula fellis, quam S1; la vésicule du fiel; et tant de la rate et du fiel, comme C1.

[85] quaedam] Quaedam etiam sanguinis portio S1; Il y en a aussi C1.

[86] renes] renes S1; la chair des rognons C1.

[87] per poros cutis.] per totam cutem emanantia. S1; au travers de toute la peau. C1.

[88] [*subscriptum:* ∧; *superscriptum:* ∧ aut] figura] aut figura S1; ou la figure C1, 9.

[89] pororum] pororum, per quos transeunt S1, 13; pores par où elles passent C1.

aliae transeant, et ne reliquum⁹⁰ sanguinis sequatur, ut videre est in diversis⁹¹ oribus quae diversimodae^xx perforata inserviunt percolendis diversis granis.⁹²

⁹³Sed quod praecipue hic notandum, magis vividae et fortiores⁹⁴ particulae sanguinis influunt concavitates⁹⁵ cerebri, quia arteriae, quae eas illuc deducunt magis omnium, quae ex corde procedunt, sitae sunt in lineae recta, et ut scis omnia corpora quae moventur, conantur quaeque quantum possunt motum suum continuare in linea recta: ⁹⁶ecce ex. gr. cor A et cogita cum sanguis vi effertur ex apertura B nullas esse eius partes quae non tendant versus C ubi sunt concavitates cerebri, sed via nimis angusta quam ut omnes eo tempore possint, debiliores a fortioribus impediuntur, quae sic solae illuc ingrediuntur.

⁹⁷Potes etiam animadvertere per transennam,^xxi nullas post eas, quae [13] cerebrum ingrediuntur, nec fortiores, nec vividiores esse illis, quae vasa generationi destinata influunt, nam ex. gr. illae quae vim habent perveniendi ad D non potentes ulterius progredi versus C quia non est satis loci pro omnibus, revertuntur⁹⁸,^xxii potius versus E quam F aut G eo quod ibi via sit magis recta.

^xxiiiPost quae possem fortasse ostendere quomodo ex humore illo qui congregatur versus C⁹⁹ possit formari alia ⟦ma⟧ machina huic prorsus similis, verum nolo in hac materia ulterius progredi.

¹⁰⁰Quod ad illas sanguinis partes quae penetrant usque ad cerebrum non tantum ibi inserviunt nutritioni et conservationi eius substantiae, sed praecipue etiam ad ibi producendum quendam ventulum admodum subtilem, aut prorsus flammam vividam, et puram admodum quam vocamus spiritus animales, sed notandum arterias, particulas illas e corde deducentes, postquam divisae fuerint, in infinitos [14] ramusculos, et compleverint parvulas illas texturas,¹⁰¹ quae instar tapetum in fundo concavitatum cerebri extenduntur, congregari⟦,⟧ circum quandam glandinem sitam fere in medio cerebri¹⁰² iuxta introitum eius concavitatum,¹⁰³ ibique infinitos prope¹⁰⁴

---

⁹⁰ ut unae potius quam aliae transeant, et ne reliquum] quod quibusdam transitus concedatur, aliis praecludatur, atque prohibeatur, ne pars reliqua S1; que les unes y passent plutôt que les autres, et que le reste C1.

⁹¹ ut videre est in diversis] Quemadmodum diversimode S1; ainsi que vous pouvez avoir vu divers C1.

⁹² granis.] generis grana S1; grains les uns des autres. C1.

⁹³ § 9 S1; § 12 C1.

⁹⁴ fortiores] vehementius actae S1; les plus fortes, et les plus subtiles C1.

⁹⁵ concavitates] concavitates sive ventriculos S1; concavités C1.

⁹⁶ § C1.

⁹⁷ § 10 S1; § 13 C1, 10.

⁹⁸ revertuntur] revertuntur S1; elles retournent C1; elles se détournent C2, 9.

⁹⁹ C] E S1, 14; E C1.

¹⁰⁰ § 14 C1.

¹⁰¹ parvulas illas texturas,] admiranda illa reticula, S1, 15; ces petits tissus, C1, 11.

¹⁰² cerebri] cerebri S1; de la substance de ce cerveau C1.

¹⁰³ concavitatum,] ventriculorum, sive concavitatum S1; concavités, C1.

¹⁰⁴ infinitos prope] infinitis fere S1; un grand nombre C1.

habent poros, per quas subtiliores sanguinis particulae, quas ⟦co⟧ continent, in glandulam istam effluere possunt, qui vero adeo sunt angusti ut crassioribus transitum non praebeant.

[105] Sciendum etiam hasce arterias ibi non finiri, sed postquam in unum congregatae fuerint, recta pergere versus amplum illud vas, quod est instar Euripi, a quo tota superficies cerebri irrigatur. Nota quoque crassiores sanguinis particulas multum posse perdere de sua agitatione in anfractibus tenuium horum contextuum,[106] per [15] quos transeunt,[107] quia ⟦quidem⟧ vim quidem habent propellendi subtiliores sibi permixtas sicque partem sui motus[108] in ipsas tranfferendi:[xxiv] hae[109] vero nequaquam possunt hoc modo perdere suam agitationem[110] quia potius augetur motu quem in ipsas transferunt crassiores, nullaque alia adsunt corpora[111] in quae tam facile transferre possent.

[112] Unde facili negotio concipere est, cum crassiores rectae[113,xxv] moventur [*in margine:* //][xxvi] versus ⟦{...}⟧ superficiem exteriorem cerebri, ubi inserviunt nutritioni substantiae, efficere ut magis subtiles et agitatas deflectantur, et ⟦{fluunt}⟧ fluant omnes in glandulam, quam imaginari debemus tanquam scaturiginem maxime abundantem, ex qua statim [*in margine:* sourdent][xxvii] undique saliunt[114,xxviii] in concavitates cerebri: et sic absque ulla alia praeparatione aut mutatione, quam quod separatae sint a crassioribus, et ad huc[115,xxix] retineant extremam illam [16] celeritatem quam illis impressit calor cordis, cessant habere formam sanguinis, ac iam spiritus animales dicuntur. [116] Simi⟦{...}⟧lac[xxx] igitur hi spiritus intrant concavitates cerebri, inde digrediuntur in poros substantiae, ex poris in nervos, ubi quatenus ingrediuntur vel etiam solummodo conantur ingredi, magis aut minus in hos quam in alios vim habent mutandi ⟦figu⟧ figuram musculorum, quibus hi nervi sunt inserti, atque hoc modo movendi omnia membra.[xxxi]

_____Ut vidisse potestis in specubus et fontibus qui reperiuntur in hortibus[117] nostrorum Regum, solam vim, qua movetur aqua, cum scatebram suam egreditur, sufficere ut commoveantur variae machinae, imo ut quibusdam instrumentis[118]

---

[105] § C1.

[106] tenuium horum contextuum,] tenuium horum reticulorum S1; des petits tissus C1.

[107] transeunt,] luctando transeunt. S1; passent; C1.

[108] partem sui motus] motus sui partem S1; la [force] C1.

[109] hae] Hae S1; ces plus petites C1.

[110] suam agitationem] impetum suum S1; la leur C1.

[111] corpora] corpora S1, 16; corps autour d'elles C1.

[112] § C1, 12.

[113] rectae] recta, S1; tout droit C1.

[114] saliunt [*in margine:* sourdent]] prosiliant S1; coulent C1.

[115] ad huc] etiamnum S1; encore C1.

[116] § 15 and beginning of part 2, C1.

[117] qui reperiuntur in hortibus] ornamenta hortorum S1; qui sont aux jardins C1.

[118] quibusdam instrumentis] instrumentis musicis S1, 17; quelques instruments C1, 13.

modulentur, vel etiam pronuncient aliquas voces iuxta diversam dispositionem[119] tubulorum per quos deducitur [{aqua}] aqua. [120]Et profecto optime assimulari possunt nervi illius machinae, quam delineo, [17] tubulis horum[121] fontium, et eius musculi ac tendines aliis organis et machi[name]na mentis[122,xxxii] motui inservientibus, spiritus vero animales aquae eas commoventi, [*in margine:* machinas][123,xxxiii] quorum cor sit origo, et cavitates speculae;[124,xxxiv] porro respiratio et aliae [*in margine: //*][xxxv] huiusmodi actiones a spiritibus[125] dependentes ipsi ordinariae et naturales, similes sunt motibus horologii aut molae, qui iugi aquas[xxxvi] fluxu continuari possunt. Obiecta externa, quae sola sua praesentia in organa sensuum agunt, atque hoc modo machinam determinant ut diversimode moveatur, quatenus partes eius cerebri sunt dispositae, recte comparantur[126] cum peregrinis, qui aliquos ex horum fontium specubus ingredientes,[127] ipsimet non cogitantes sunt causa motuum qui coram ipsis fiunt: non possunt enim eo intrare quin incedant super quaedam pavimenti quadra sic disposita, ut si ex. gr. propius accedant ad Dianam qui se baigne[128,xxxvii] efficient, ut inta[xxxviii] arundines se recundat,[xxxix] et si [a]ulterius pergant ac illam prosequantur, aderit qui[18]dam[129] Neptunus suo tridente iis minitans, vel si aliorsum recedant versus eos egreditur quoddam monstrum marinum ipsis in faciem aquam eructans, aut simile quid, prout industrius fuerit artifex,[130 xl]

Cum autem anima rationalis huic machinae inerit, praecipuam suam sedem habebit in cerebro eritque ibi tanquam ille qui curam fonticularum gerit, qui debet vice speculis quo deferuntur omnes tubuli dictarum machinarum, ut nempe pro ut libuerit exitet, impediat, aut mutet aliquo modo earum [mod] motus.

[131]Sed ut tandem vobis haec omnia distinct[a]e demonstrem loquar primum de structura nervorum et musculorum, atque ostendam quomodo ex eo solummodo quod spiritus qui insunt [*in margine: #*][xli] cerebro nervum quendam ingredi conentur, vim habeant illico movendi aliquod membrum: postea ubi verbo attigero respirationem et alios eiusmodi motus simplices et ordinarios, dicam quomodo [19] obiecta externa

---

[119] dispositionem] varietatem S1; disposition C1.
[120] § 16 C1.
[121] horum] horum S1; des machines de ces C1.
[122] organis et machi[name]na mentis] organis, et instrumentis S1; engins et ressorts C1.
[123] eas commoventi, [*in margine:* machinas]] ea movet concitatque. S1; les remue, C1.
[124] et cavitates speculae;] concavitates vero cerebri, castellum, sive aquarum receptacula. S1; et les concavités du cerveau sont les regards. C1; et dont les concavités du cerveau sont les regards. C2, 12.
[125] a spiritibus] a spiritibus S1; du cours des esprits. C1.
[126] recte comparantur] recte comparantur S1; sont comme C1.
[127] qui aliquos ex horum fontium specubus ingredientes,] horum fontium rupiumve spectatoribus; qui forte istuc ingressi, S1; qui entrant dans quelques-unes des grottes de ces fontaines, C1.
[128] qui se baigne] quae forte se lavat S1; qui se baigne C1.
[129] aderit quidam] prodeat forte S1, 18; ils feront venir vers eux un C1.
[130] prout industrius fuerit artifex,] pro artificum industria S1; selon le caprice des ingénieurs qui les ont faites; C1.
[131] § 17 C1, 14.

agant in sensuum organa, ac denique speciatim explicabo omne id quod peragitur in concavitatibus et poris cerebri, quo[132] spiritus animales tendant,[133] et quaenam sint nostrae functiones quas machina earum ope representare potest, si enim inciperem a cerebro, atqui ordine tantum sequerem cursum spirituum, ut sanguinis ⟦cursus se⟧ cursum secutus sum videor mihi non tam facile posse intelligi.[134]

[135]Ecce itaque hic ex. gr. nervum A cuius exterior tunica similis est amplo [*in margine:* gallice pellis][136,xlii] satis tubo, qui in se continet plures alios exiguos tubulos[xliii] B C K L[137] constantes tunica interiori magis tenui:[138,xliv] [*in margine:* //][xlv] quae ambae[139] continuae sunt duabus aliis[140] K L quibus involvitur cerebrum MNO. [141]Nota etiam in unoquoque horum exiguorum tubulorum aliquod esse instar medullae constanti[ccxl] ex pluribus filamentis tenuissimis, quae proveniunt ex ipsa substantia cerebri N et quorum extremitates terminantur ab una parte in eius superficie interiori, quae ipsius concavitates respiciunt, et ab alia parte in tuni[20]cis [*in margine:* //][xlvi] et carne inquam[xlvii] tubulus[142,xlviii] qui illa[143,xlix] continet, finitur. Verum quoniam haec medulla[144] nullo modo inservit motui membrorum, mihi hoc tempore sufficit si scias ipsam non sic implere exiguos illos tubulos, qui illam continent, quin adhuc detur satis spatii ut spiritus animales quam facillime a cerebro in musculos fluere possint, quo ⟦ipsi⟧ [*superscriptum:* isti][145,l] tubuli[li] qui totidem sunt censendi nervi tendunt. [146]Nota postea quo pacto nervulus[147] BF ingrediatur musculum D quem suppono esse unum ex illis qui oculum movent, et quomodo postquam eo devenerit dividantur[ccxli] in plurimos ramusculos constantes tunica flaccida, quae distendi aut dilatari et contrahi posset, prout multi fuerint spiritus animales, qui eo ingrediuntur, aut ex illo egrediuntur et quorum fibrae[148] ita sunt constitutae

---

[132] quo] quo S1; comment C1.

[133] tendant,] tendunt, S1; prennent leurs cours; C1.

[134] videor mihi non tam facile posse intelligi.] caetera minus clare perciperentur. S1, 19; il me semble que mon discours ne pourrait pas être du tout si clair. C1.

[135] § 18 C1, 15.

[136] tunica [*in margine:* gallice pellis]] membrana, sive tunica S1; la peau C1.

[137] B C K L] b. c. k. l. etc. S1; b, c, k, l, etc. C1.

[138] magis tenui:] et tenuiori S1; plus déliée; C1.

[139] ambae] ambae tunicae S1; deux peaux C1.

[140] duabus aliis] duabus S1, 20; les deux C1.

[141] § C1.

[142] et carne inquam tubulus] et carnem, in quibus tubulus S1; et aux chairs contre lesquelles le tuyau C1.

[143] illa] filamenta illa S1, 21; les C1.

[144] haec medulla] medullam istam S1; elle C1.

[145] ⟦ipsi⟧ [*superscriptum:* isti]] isti S1; ces C1.

[146] § 19 C1, 16.

[147] nervulus] nervulus S1; le tuyau, ou petit nerf C1.

[148] fibrae] fibrae S1; les rameaux ou les fibres C1, 17.

ut cum spiritus[149] introierunt efficiant ut totus musculus[150] intumescat atque contrahatur, ⟦cum ingrediuntur⟧ et sic trahat oculum cui annexus est. [21] Sicut contra cum egrediuntur,[151] tunc musculus ille detumescit ⟦ext⟧ et extenditur.[152,lii]

[153]Praeterea vide praeter tubum BF unum esse alium nempe EF per quem spiritus animales ingredi possunt musculum D et alterum sc. DG per quos inde egredi possunt, atque eodem modo musculum E quem suppono inservire movendo oculo, in partem contrariam[154] recipere spiritus animales a cerebro per tubum CG et a musculo D per G[155] eosque iterum remitti versus D per EF, et cogita licet nullus evidens sit meatus, per quem spiritus contenti in duobus musculis D E[156] inde egredi possint, nisi ut ex uno in alium ingredia[n]tur, tamen quia particulae eius^{ccxlii} admodum sunt exiguae, imo etiam subtilisantur magis magisque vi illarum agitationis semper aliqui elabuntur per pelliculas et horum musculorum carnes, verum deperditorum loco semper aliqui veniunt per duos tubulos BF CG. [*in margine:* ⟦hf⟧][157,liii]

[158]Denique nota certam quandam [22] pelliculam esse[159] HF[160] quae separat duos tubulos BF EF[161] quaeque [*in margine:* H̲ I̲]^{liv} duas habet plicas H̲ ⟦F⟧I[162] sic dispositas ut cum spiritus animales qui conantur ⟦des⟧ descendere a B versus H maiorem habent impetum quam ii, ⟦as⟧ qui ascendere nituntur [*in margine:* I̲]^{lv} ab E versus ⟦F⟧I[163] deprimant et aperiant hanc pelliculam,[164] sicque praebeant occasionem illis qui contenti sunt in musculo E ut promptissime fluant secum versus D, sed cum illi qui conantur ascendere ab E versus I fortiores sunt, aut solummodo cum sunt aeque fortes ac alii, attollunt et claudunt pelliculam HF[165] et ita se ipsos impediunt quo minus exeant ex musculo E, cum contra si utrimque satis virium habeant[166] ut per illam transeant,[167,lvi] naturaliter aperta manet: et denique si aliquando spiritus

---

[149] spiritus] spiritibus S1; esprits animaux C1.

[150] totus musculus] musculus S1; tout le corps du muscle C1.

[151] egrediuntur,] spiritibus egredientibus S1; en ressortent C1.

[152] extenditur.] laxetur, et extendatur. S1; se rallonge. C1.

[153] § 20 C1.

[154] in partem contrariam] in adversam partem S1, 22; tout au contraire du précédent C1.

[155] G] G S1; dg C1.

[156] D E] D E S1; D et E C1.

[157] BF CG. [*in margine:* ⟦hf⟧]] B F et C G eo defluentes. S1; bf cg. C1, 18.

[158] § 21 C1.

[159] esse] dari S1; qu'entre les deux tuyaux bf, e f, il y a C1.

[160] HF] HFI S1; Hfi C1.

[161] duos tubulos BF EF] duos tubulos BF et EF S1; ces deux tuyaux, et qui leur sert comme de porte C1.

[162] plicas H̲ ⟦F⟧I [*in margine:* H̲ I̲]] plicas, sive valvulas H I S1; deux replis H et i C1, 18–19.

[163] ⟦F⟧I [*in margine:* I̲]] I S1; i C1, 19.

[164] pelliculam,] valvulam sive pelliculam: S1; peau C1.

[165] pelliculam HF] pelliculam sive valvulam HFI S1, 23; peau Hfi C1.

[166] habeant] non habeant S1; n'ont pas C1.

[167] per illam transeant,] per illam transeant, S1; pour la pousser, C1.

contenti in musculo D inde digredi satagant per DFE aut DFB plica[168] H distendi potest et illis occludere viam. Similiter intra duos tubulos CG DG[169] [23] quaedam pellicula[170] est similis praecedenti, quae occludi potest[171] a spiritibus venientibus a tubulo DG et aperitur ab iis qui veniunt a CG [172]postquae[173,lvii] [*in margine: //*][lviii] facile erit intellectu si spiritus animales qui sunt in cerebro fluere non conentur aut fere non,[174] per tubulos BF CG duas istas pelliculas[175] F et G reclusas[176] mansuras, quin etiam duos musculos D E[177] flaccidos et absque actione fore, quia spiritus animales quos continent libere ex uno in alterum fluunt cursum suum orsi[178,lix] ab E [*in margine: //*][lx] per F versus D et reciproce a D per G versus E. Verum e[nim] v[ero] si spiritus qui in cerebro sunt contenti quodam impetum nitantur ingredi duos tubos BF CG atque iste impetus utrumque sit aequalis, occludunt e vestigio ambos meatus G F[179] et inflant duos musculos D E quantum possint, quo fit ut sistatur oculus et firme maneat in eodem situ qua inventus fuit. [180]Deinde si spiritus qui ex cerebro proveniunt [24] maiori impetu fluere conantur per BF quam EGI[181] occludunt pelliculam[182] G, [[recul]] recludunt F,[183] idque minus aut magis celeriter prout pellicula F[184] magis aut minus aperta est, ita ut musculus D unde hi spiritus egredi nequeunt contrahitur, et E extenditur, quo fit ut oculus vertatur versus D sicut contra si spiritus qui sunt in cerebro maiori impetu conentur fluere per CG quam per BF claudunt pelliculam F et aperiunt G ita ut spiritus musculi D redeant statim in musculum[185] E, quo fit ut hic contrahatur et oculum in suam partem trahat. [186]Hi enim[187] spiritus cum sint instar ventuli aut flammae subtilissimae, non possunt quin [[s]] citissime fluant [[u]]ex

---

[168] plica] valvula S1; le repli C1.

[169] CG DG] C G et D G S1; cg, dg, C1.

[170] pellicula] valvula S1; une petite peau ou valvule g C1.

[171] quae occludi potest] quae occludi potest S1; qui demeure naturellement entr'ouverte, et qui peut être fermée C1.

[172] § C1.

[173] postquae] Quibus explicatis S1; En suite de quoi C1.

[174] fluere non conentur aut fere non,] non tendant, aut fere non conentur fluere S1; ne tendent point, ou presque point, à couler C1.

[175] pelliculas] valvulas S1; les deux petites peaux ou valvules C1.

[176] reclusas] semiapertas sive hiantes S1; entr'ouvertes C1.

[177] D E] D E S1; D et E C1.

[178] cursum suum orsi] exorsi S1; prenant leur cours C1.

[179] G F] G F S1; g et f C1, 20.

[180] § C1, 21.

[181] EGI] CG S1, 24; cg C1.

[182] pelliculam] pelliculam, sive valvulam S1; la petite peau C1.

[183] F,] F; S1; f; et ce plus ou moins, selon qu'ils agissent plus ou moins fort; au moyen de quoi les esprits contenus dans le muscle E se vont rendre dans le muscle D, par le canal ef C1.

[184] pellicula F] ea valvula S1; peau f C1.

[185] in musculum] in musculum S1; par le canal dg dans le muscle C1.

[186] § C1.

[187] Hi enim] Manifestum enim est hos S1; Car vous savez bien que ces C1.

uno musculo in alterum simula⟦{…}⟧c quosdam meatus inveniunt, licet nulla alia sit vis quae eos illuc deducat quam sola inclinatio[188] continuandi sui motus iuxta leges naturae, et quanquam[189] ⟦sit⟧ sint valde mobiles et subtiles, non obstat quo [25] minus vim habeant inflandi et distendendi[190] musculos quibus sunt inclusi, quemadmodum aer in folliculo contentus ipsum inducat[191] ac pelles intendit,[192] quibus includitur, [193]itaque[194] facile e⟦x⟧st applicare ⟦{non} applicare⟧[195] id quod dico de nervo A et duobus aliis musculis D E[196] aliis omnibus musculis et nervis, et hoc pacto intelligere[197] quomodo machina, de qua loquor, possit moveri tam diversimode ac nostra corpora solo impetu spirituum animalium, qui fluunt a cerebro in nervos, nam pro unoquoque motu eiusque contrario[198] potes imaginari duos tubos tales[199] quales sunt BFD CGE[200] et duos alios quales DG EF[201] ac duas exiguas valvulas[202] quales fuere HF⟦E⟧I[lxi] [*in margine: //*][lxii] [203] et quod ad modos quibus hi tubuli musculis inseruntur, etiamsi in mille formas[204] varient, tamen non difficile erit iudicare quinam illi sunt, si scia[s] quod anatomia[lxiii] docere potest figuram externam, et usum cuiusque musculi.

[205]Nam sciens ex. gr. palpebras moveri musculis,[206] quorum unus nempe T nulli alii rei inservit quam ⟦{…}⟧ [26] aperiendae superiori, et alter scil. V inservit alternatim eas aperiendi et ⟦clad⟧ claudendi ambas, facili negotio cogitabis ipsas recipere spiritus per duos tubulos tales, quales sunt PV[207] et QS et unum ex his ⟦{…}⟧PV[208] tendere versus duos ill⟦{…}⟧os musculos, et alterum QS versus unum illorum tantum, et denique ramusculos R A G[209,lxiv] fere eodem modo musculo V insertos. Duo

---

[188] inclinatio] conatum S1; inclination C1.

[189] et quanquam] Neminem quoque latet, quanquam S1; Et vous savez outre cela, qu'encore C1.

[190] inflandi et distendendi] inflandi, distendendi atque hac ratione indurandi S1; d'enfler et de raidir C1.

[191] ipsum inducat] tumidiorem illam durioremque efficit S1, 24–25; le durcit C1.

[192] intendit,] distendere solet. S1; fait tendre C1.

[193] § 22 C1, 22.

[194] itaque] igitur cognitu S1, 25; Or C1.

[195] applicare ⟦{non} applicare⟧] quadrare S1; appliquer C1.

[196] D E] D E S1; D et E C1.

[197] et hoc pacto intelligere] Neque intellectu difficilius est S; et ainsi d'entendre C1.

[198] contrario] contrariae determinationi S1; contraire C1.

[199] tubos tales] tubos tales operam navare S1; petits nerfs, ou tuyaux, tels C1.

[200] BFD CGE] B F D et C G E S1; bf, cg, C1.

[201] DG EF] D G et E F S1; dg, ef, C1.

[202] valvulas] valvulas S1; petites portes ou valvules C1.

[203] § C1.

[204] mille formas] millies S1; en mille sortes C1.

[205] § 23 C1.

[206] musculis,] musculis, S1, 26; deux muscles, C1, 23.

[207] PV] P V S1; pR C1.

[208] PV] P V S1; pR C1.

[209] R A G] r et g S1; R et S C1.

tamen effecta sortiuntur omnino contraria ob diversam fibrarum²¹⁰ dispositionem, quod sufficit ad alios motus intelligendos. ²¹¹Imo non difficulter poteris iudicare ex eo quod spiritus animales in omnibus membris aliquos motus ciere possint ibi etiam aliquos terminari nervos, etiamsi plurima sint ubi anatomici nullos visibiles animadvertant, ut in pupilla,²¹² in corde, iecore, felle, liene et eiusmodi aliis.ˡˣᵛ

[27] ²¹³Ut iam specialius intelligamus, quomodo haec machina respiret cogitandum musculum D ex eorum numero esse qui inserviunt pectori attollendo aut deprimendo diaphragmati, eique contrarium esse musculum E, et spiritus animales qui sunt in concavitate cerebri notata M²¹⁴ fluentes per porum aut exiguum canalem²¹⁵ N qui naturaliter²¹⁶ semper apertus manet, statim differri in tubum BF deprimendo pelliculam F quo fit ut illi²¹⁷ qui continentur in musculo E inde progrediendo inflent²¹⁸ musculum D. ²¹⁹Cogita etiam quasdam esse pelliculas circumcirca musculum D quae eum magis magisque prem[[{...}]]unt <u>simulac exspirat</u>²²⁰,ˡˣᵛⁱ sicque sunt dispositae ut antequam omnes spiritus musculi E versus eum digressi fuerint, cursum suum sufflaminent, et quasi egurgitentur per tubulum BF ita ut illi qui sunt in canali MN²²¹ inde deflectantur, quo mediante [*in margine:* <u>NB</u>]ˡˣᵛⁱⁱ in tubum <u>CG</u>ˡˣᵛⁱⁱⁱ delati,²²² quaeᶜᶜˣˡⁱⁱⁱ ape[28]riunt eodem tempore, musculum E tumescer[e] faciunt, ac detumescere²²³ musculum D quod tam diu continuant quam diu durat impetus quo spiritus contenti in musculo D pressi a pelliculis eum circumdantibus, inde aegrediˡˣⁱˣ nituntur, cum autem impetus vim amisit²²⁴ sponte sua rursus eundem cursum repetunt per tubulum [[{...}]] B²²⁵ et sic non cessant alternatim intumescere et detumescere hi duo musculi. Quod idem iudicandum de aliis musculis eidem effecto inservientibus, et cogita²²⁶ omnes ita esse dispositos ut cum illi, qui similes sunt musculo D,²²⁷ intumescunt, spatium, quo continentur pulmones, dilatetur, quae causa est quod

---

²¹⁰ fibrarum] fibrarum S1; de leurs rameaux ou de leurs fibres; C1.

²¹¹ § C1.

²¹² pupilla,] pupilla, S1; la prunelle de l'œil, C1.

²¹³ § 11 S1; § 24 C1.

²¹⁴ notata M] M S1, 28; marqué m C1.

²¹⁵ porum aut exiguum canalem] exiguum canalem S1; pore ou petit canal marqué C1.

²¹⁶ naturaliter] secundum naturae requisitum dispositus S1; naturellement C1.

²¹⁷ illi] spiritus S1; ils C1.

²¹⁸ inde progrediendo inflent] inde progredientes flatu suo distandant S1; viennent enfler C1.

²¹⁹ § C1, 24.

²²⁰ <u>simulac exspirat</u>] prout uberius inflatur S1; à mesure qu'il s'enfle C1.

²²¹ MN] M N S1; n C1.

²²² [*in margine:* <u>NB</u>] in tubum <u>CG</u> delati,] in tubum C G (...) sese conferentes, S1; s'allant rendre dans le tuyau cg C1.

²²³ detumescere] evacuant S1; désenfler C1, 25.

²²⁴ vim amisit] non amplius tantae est efficaciae S1; n'a plus de force C1.

²²⁵ B] B S1; BF C1.

²²⁶ et cogita] Unde constat S1, 28–29; et penser C1.

²²⁷ musculo D,] musculo D, S1, 29; à d C1.

aer ipsum subeat eodem modo ut in follem cuius latera diducuntur:[228] si autem iis contrarii intumescant, spatium hoc coarctetur, quo fit ut egrediatur aer.

[229]Ut etiam intelligamus quomodo haec machina cibos deglutiat qui reperiuntur in fundo oris cogitandum [29] [[{D}]] musculum D esse unum ex illis qui radicem linguae attollunt et meatum[230] apertum detinent, per quem aer quem respirat pulmones ingredi[231] debet, ac musculum E esse ei contrarium, qui inservit occludendo huic meatui et eadem ope illi[232] aperiendo, per quem cibos qui in ore sunt in stomachum descendere oportet, vel etiam ut attollat apicem linguae, qui ipsos eo versus pellat, nec non spiritus animales, qui e concavitate cerebri M fluunt per porum aut canalem exiguum N qui naturaliter semper apertus est, omnes recta deferri in tubum BF quo fit intumescat musculus D et denique semper sic tumidum manere dum nulli reperiuntur cibi in fundo oris, qui eum premant. Verum sic est constitutus, ut simulac aliqui reperiantur spiritus quos continet, illico egurgitet per tubum BF et efficiat ut illi qui veniunt per canalem N musculum E per tubum CG ingrediantur [30] quo etiam tendunt illi qui sunt in musculo D et sic gula aperitur, ac cibi in stomachum descendunt: tunc paulo post spiritus canalis N cursum suum ut ante per BF dirigunt, [233]ad cuius similitudinem potest etiam intelligi quo pacto haec machina sternutet, oscitet, tussiat, et varios alios peragat motus necessarios, ut diversa eiiciantur excrementa.

[234]Ut intelligamus postea quomodo incitari possit ab obiectis externis quae afficiunt sensuum organa, ut omnia sua membra infinitis modis[235] moveat cogitandum exigua filamenta, quae modo dixi provenire ex interiori ipsius cerebri[236] et componere[237,lxx] medullam eius nervorum, inesse[238] omnibus suis partibus, sensu alicuius organo inservientibus, ibidem[239] facillime posse moveri a sensuum horum obiectis, et cum paululum moventur ipsa e vestigio trahere partes cerebri e quibus proveniunt, et aperire eadem [31] opera meatus quorundam pororum qui sunt in superficie cerebri,[240] unde spiritus animales[241] statim suum cursum orsi per eosdem in nervos et musculos deferuntur qui inserviunt excitandis[242] motibus omnino similibus iis quibus naturaliter incitamur cum nostri sensus eodem modo afficiuntur [243]ut ex.

---

[228] cuius latera diducuntur:] deductis lateribus S1; que l'on ouvre; C1.
[229] § 12 S1; § 25 C1.
[230] meatum] asperae arteriae orificium S1; le passage C1.
[231] pulmones ingredi] ingrediatur, S1, 31; entrer dans son poumon, C1.
[232] illi] oesophagi orificium S1; celui C1.
[233] § C1, 26.
[234] § 13 S1, 31; § 26 C1.
[235] infinitis modis] mille modis S1; mille autres façons C1.
[236] interiori ipsius cerebri] intimis ipsius cerebri penetralibus S1; plus intérieur de son cerveau C1.
[237] componere] componere S1; composer C1.
[238] inesse] inesse, S1; sont tellement disposés en C1.
[239] ibidem] atque haec quidem filamenta S1; qu'ils C1.
[240] cerebri,] interiori cerebri S1; intérieure de ce cerveau, C1.
[241] animales] animales S1; animaux qui sont dans ses concavités C1.
[242] inserviunt excitandis] excitant S1; servent à faire en cette machine C1, 26–27.
[243] § C1, 27.

gr. si ignis A sit prope pedem B particulae huius ignis celerrime[244] motae vim habent secum movendi partem pellis huius pedis quem attingunt, et sic trahentes funiculum CC quem ibi videtis[245] annexum eodem momento aperiunt meatum pori DE contra quem hic funiculus terminatur quemadmodum si trahas extremitatem funis eodem momento efficis ut sonet campana quae ab alia extremitate pendet: [246]meatu igitur sive introitu[247] pori aut exigui ductus D[248] hoc modo aperto, spiritus animales concavitatis F in eundem influunt et ab eo deducuntur, partim in musculos qui inserviunt retrahendo pedi ab hoc igne, partim in eos qui inserviunt obvertendo[249] oculos et caput ad eum respiciendum et in illos qui inserviunt porrigendis manibus et toto corpori in[32]flectendo ut ipsum defendamus,[250,lxxi] [251]verum deferri quoque possunt per hunc ipsum ductum DE[252] in plures alios musculos. Et antequam insistam exactius explicare[253] quo pacto spiritus animales cursum suum sequantur per poros cerebri, et quomodo hi spiritus[254] sint dispositi, loquar hic primum speciatim de omnibus sensibus quemadmodum in hac machina reperiuntur ac dicam quomodo ad nostros referantur.

Tactus[lxxii]

[255]Scias igitur primo ingentem esse numerum exiguorum filamentorum similium CC quae omnia incipiunt ab invicem separari a superficie interiori cerebri, unde suum[lxxiii] originem sortiuntur, quae inde dispersa per totum[256] corpus inserviunt organis tactus,[257] etiamsi enim ut plurimum ipsa non sint quae immediate tanguntur ab obiectis externis, sed cutes [*superscriptum:* pellicul[{...}]ae][258,lxxiv] quae ea circumdant, nullatenus tam tamen verisimilius[259] est has ipsas cutes esse organa sensus, quam cum aliquod corpus contractamus chirotecis induti, has

---

[244] celerrime] uti constat, celerrime motae S1, 32; comme vous savez très promptement C1.

[245] ibi videtis] illi videtis S1; vous voyez y C1.

[246] § C1, 28.

[247] meatu igitur sive introitu] Orificio igitur S1; Or l'entrée C1.

[248] D] D S1; d, e, C1.

[249] inserviunt obvertendo] obvertat S1; servent à tourner C1.

[250] et in illos qui inserviunt porrigendis manibus et toto corpori inflectendo ut ipsum defendamus,] corpusque ad sui ipsius defensionem inflectat. S1; et partie en ceux qui servent à avancer les mains et à plier tout le corps pour le défendre C1; et partie en ceux qui servent à avancer les mains et à plier tout le corps pour y apporter du secours. C2, 26.

[251] § C1.

[252] per hunc ipsum ductum DE] spiritus animales ita excitati per hunc ductum D E, non tantum in hos musculos (...) sed S1; par ce même conduit d, e, C1.

[253] insistam exactius explicare] ex professo me tradam exactiori explicationi, qua manifestum fiat S1; je m'arrête à vous expliquer plus exactement C1.

[254] spiritus] pori S1; pores C1.

[255] § 27 and beginning of part 3, C1.

[256] totum] totum S1, 34; tout le reste de C1.

[257] tactus,] tactus. S1; pour le sens de l'attouchement. C1, 28–29.

[258] cutes [*superscriptum:* pellicul[{...}]ae]] membranulas S1; les peaux C1, 29.

[259] verisimilius] verisimilius S1; plus d'apparence de penser C1.

chirote⟦{...}⟧cas sensui {inservir[i]}. ²⁶⁰Et nota [33] licet filamenta, de quibus loquor, admodum sint tenuia, non eo minus tuto a cerebro usque ad alia membra magis remota perveniunt, cum interea^lxxv nihil²⁶¹ inveniant quo rumpantur aut illorum actio impediatur ipsa premendo, quamquam hae membra interim infinitis²⁶² modis f⟦{...}⟧lectantur quoniam sunt inclusa in iisdem tubulis, qui spiritus animales in musculos deducunt, hique spiritus paululum semper hosce tubulos inflantes ⟦olbs⟧ obstant, quo minus ibidem magis premantur, quin etiam ipsa continuo²⁶³ quantum possunt expandunt ⟦a⟧inde a cerebro unde procedunt versus loca ubi finiuntur.

²⁶⁴Dicam itaque postquam Deus Opt. Max. animam rationalem huic machinae univerit (ut ostendam postea) praecipuam ei adsignavit sedem in cerebro, eamque eius faciet naturae, ut iuxta diversos modos, quibus introitus, pororum, in superficie interioris cerebri sitorum⟦,⟧ aperti fuerint, mediantibus nervis, varios etiam habitura sit sensus: ²⁶⁵quemadmodum primo si exigua filamenta nervorum medullam componentia, tanto impetu trahantur ut [34] disrumpantur, et a parte cui iuncta ⟦era⟧ erant separentur, ita ut totius machinae structura ex eo aliquoties fiat minus perfecta,²⁶⁶ motus quem in cerebro quem in cerebro [*sic*] causabuntur, dabit animae occasionem²⁶⁷ (cuius interest ut sedes sua²⁶⁸ conservetur) dolorem sentiendi. ²⁶⁹Et si trahantur impetu fere aequali praecedenti²⁷⁰ sic tamen ut non rumpantur, nec a partibus quibus annectuntur separentur, efficient motum in cerebro, qui testimonium praebens bonae aliorum membrorum constitutionis, dabit occasionem²⁷¹ animae sentiendi quandam voluptatem corpoream quam titillationem vocamus, quaeque dolori, quoad ad causam suam, proxima ipsi contrarium omnino sortitur effectum, ²⁷²quod si plura huiusmodi filamenta simul aequaliter trahantur efficient ut anima sentiat superficiem corporis, quod tangit membrum in quo finiuntur, esse politam:²⁷³ efficient contra ut sentiat illam esse inaequalem et asperam²⁷⁴ ⟦inae⟧ si inaequaliter

---

²⁶⁰ § C1.
²⁶¹ interea nihil] nihil S1; rien entre deux C1.
²⁶² infinitis] infinitis S1; en mille diverses C1.
²⁶³ quin etiam ipsa continuo] et simul ut S1, 35; et même qu'ils les font toujours C1.
²⁶⁴ § 28 C1.
²⁶⁵ § 29 C1.
²⁶⁶ fiat minus perfecta,] corrumpitur, S1; en soit en quelque façon moins accomplie, C1.
²⁶⁷ dabit animae occasionem] anima (...) causam concipiet S1; donnera occasion à l'âme, C1.
²⁶⁸ sedes sua] sedem suam sartam tectam S1; le lieu de sa demeure C1.
²⁶⁹ § C1, 30.
²⁷⁰ aequali praecedenti] pari S1; aussi grande que la précédente C1.
²⁷¹ dabit occasionem] dabit causam S1, 36; donnera occasion C1.
²⁷² § 30 C1.
²⁷³ politam:] laevem atque politam. S1; polie; C1.
²⁷⁴ illam esse inaequalem et asperam] asperam superficiem significabunt S1; polie; inégale, et qu'elle est rude C1.

trahantur. ²⁷⁵Si autem aliquantulum separatim²⁷⁶ impellantur, ut continuo fit a calore quem cor aliis [37]^lxxvi membris communicat nullum inde²⁷⁷ anima percipiet sensum, non magis quam aliarum actionum quae ipsi sunt ordinariae; verum si hic motus in ipsis aliquanto magis augeatur vel diminuatur ab aliqua causa externa,²⁷⁸ ipsius augmentatio praebebit animae sensum caloris, diminutio frigoris. Et denique iuxta varios modos quibus movebuntur²⁷⁹ efficient ut sentiat omnes alias qualitates ad tactum in genere pertinentes, ut humiditatem, siccitatem, duritiem, gravitatem,²⁸⁰ et similes.²⁸¹Solummodo hic notandum²⁸² licet admodum sint tenuia et²⁸³ moveri facile possint non tamen id sunt^lxxvii tantopere²⁸⁴ ut cerebro omnium [*in margine: //*]^lxxviii minimas actiones²⁸⁵ referre queant, sed minimas earum²⁸⁶ quae cerebro referuntur esse eas quas afficiunt²⁸⁷ corporum terrestrium particulae crassiores, imo horum corporum²⁸⁸ aliqua{ [[e]] } esse posse quorum particulae sint satis crassae, nihilominus tamen adeo leniter in haec exigua filamenta irrepent, ut ea pressurae vel etiam prorsus scissurae sint, sic ut earum actio usque ad cerebrum non transeat, quemadmodum quaedam [38] sunt pharmaca quae vim habent torporem inducendi, imo et corrumpendi actiones nostrorum membrorum quibus applicantur, ita ut ne quidem sentiamus.

Gustus^lxxix

²⁸⁹Verum exigua filamenta ex quibus constat medulla nervorum linguae, quaeque inserviunt organis gustus²⁹⁰ moveri possunt minoribus actionibus,²⁹¹ quam ea quae inserviunt tactui in genere, tum quia paulo sunt tenuiora, tunc etiam quia pelliculae quibus involvuntur sunt teneriores. ²⁹²Cogita itaque illa multifariam²⁹³ posse moveri

---

²⁷⁵ § C1.

²⁷⁶ aliquantulum separatim] non simul, sed alia post alia quadantenus separatim S1; quelque peu séparément l'un de l'autre C1.

²⁷⁷ nullum inde] nullum omnino S1; n'en (...) aucun C1.

²⁷⁸ externa,] externa S1; extraordinaire, C1.

²⁷⁹ movebuntur] nervorum filamenta afficiuntur, anima S1; ils seront mus, ils lui C1.

²⁸⁰ duritiem, gravitatem,] duritiem, gravitatem S1; la pesanteur, C1, 31.

²⁸¹ § 31 C1.

²⁸² notandum] notandum, quamvis illa S1; remarquer qu'encore qu'ils C1.

²⁸³ et] adeoque S1; et C1.

²⁸⁴ non tamen id sunt tantopere] non tamen tam exilia esse S1; ne le sont pas toutefois tellement C1.

²⁸⁵ omnium minimas actiones] omnium minimos motus S1, 37; toutes les plus petites actions qui soient en la nature C1.

²⁸⁶ minimas earum] omnium minimos motus S1; les moindres qu'ils C1.

²⁸⁷ quas afficiunt] quos (...) causantur S1; celles des C1.

²⁸⁸ horum corporum] istius generis S1; de ces corps C1.

²⁸⁹ § 14 S1; § 32 C1.

²⁹⁰ gustus] gustus S1; le goût en cette machine C1.

²⁹¹ minoribus actionibus,] facilius S1; par de moindres actions, C1.

²⁹² § C1.

²⁹³ multifariam] diversis quatuor modis S1; en quatre diverses façons C1.

a particulis ex. gr. salis, aquae acidae, communis, vitae, quarum[294] supra quantitatem et figuras explicui,[295,lxxx] et hoc modo efficere ut sentiat animam quatuor sapores omnino[296] differentes, quoniam salis particulae agitatae et ab invicem separate salivae, motu, linguae[297] poros punctim et nequaquam inflexae ingrediuntur. Aquae autem acidae particulae eosdem influunt obliqu[[{a}]]e, [*in margine:* #][lxxxi] partes tenerrimas[298,lxxxii] et crassioribus obe[39]dientes[299] incidendo.[300] Eae vero quibus constat aqua communis nullas illius partes incidentes neque etiam alte poros ingredientes supra ipsam solummodo molliter fluitant: ac denique aquae vitae[301] particulae omnium altissime [[{...}]] penetrant motuque celerrimo concitantur. Unde facile iudicabis quo pacto anima caeteras[302] corporum[303] species sentiat, si consideres quam diversimode corporum terrestrium particulae in linguam agere possint. [304]Sed quod maxime notandum, eaedem sunt ciborum partes, et quae dum in ore sunt linguae poros ingredi, ibique gustus sensum ciere possunt, et quae e stomacho effluentes in sanguinem mutantur ac inde reliquis membris uniuntur,[305] quia etiam et illae tantum, quae linguam leniter titillant, quaeque hoc pacto animae gratum saporis sensum praebent, huic rei[306] omnino sunt idoneae. [307]Nam quod ad eas quae nimium aut parum in ipsam agunt, ut sunt nimis pungentes, aut insipidae nimis,[308] sic etiam sunt nimium aut penetrantes, aut molles, quam [40] ut sanguinem componant, ac membra quaedam alant: illae autem[309] quae sunt adeo crassae sive tantopere unitae ut salivae, motu separari[310] non possunt, nec ullo modo penetrare in poros linguae, ut nempe agant exigua capillamenta nervorum gustui inservientium, aliter quam eos nervos quibus in reliquis membris tactus[311] peragit, quique etiam ipsi nullos poros habent, per quos linguae particulae, vel ad minimum salivae, quibus humectatur

---

[294] quarum] quarum omnium particularum S1; dont C1.

[295] supra (…) explicui,] tibi explicui: S1; je vous ai ci-dessus expliqué C1.

[296] sapores omnino] sapores omnino S1; sortes de goûts C1, 32.

[297] linguae] linguae S1, 38; la peau de la langue C1.

[298] partes tenerrimas] tenuissimas particulas S1; les plus tendres de ses parties C1.

[299] et crassioribus obedientes] quae tamen in crassiores impingentes inflectuntur S1; et obéissant aux plus grossières C1.

[300] incidendo.] incidendo: S1; en tranchant ou incisant C1.

[301] aquae vitae] aquae vitae S1; de l'eau-de-vie étant fort petites C1.

[302] caeteras] caeteras S1; toutes les autres C1.

[303] corporum] saporum S1; de goûts C1.

[304] § 33 C1.

[305] uniuntur,] uniri. S1; s'aller joindre et unir C1.

[306] rei] effectui S1; effet C1.

[307] § C1.

[308] ut sunt nimis pungentes, aut insipidae nimis,] gustum excitare nequeunt nisi nimis acrem, vel nimis imbellem, S1; comme elles ne sauraient faire sentir qu'un goût trop piquant, ou trop fade, C1.

[309] illae autem] Illae vero cibi particulae, S1, 39; Et pour celles C1, 33.

[310] separari] separari, atque dilui S1; être séparées C1.

[311] tactus] tactum in genere sic dictum S1; l'attouchement en général, C1.

ingredi possunt, quemadmodum enim non poterunt efficere ut anima ullum sentiat saporem,³¹² sic quoque ineptae ut plurimum erunt quae stomacho immutantur.

³¹³Hocque adeo universaliter verum est, ut saepe quotiescunque mutatur temperamentum [*in margine:* gustus diminuetur]³¹⁴,lxxxiii etiam 〚mutetur vis saporis〛 adeo ut cibus qui consueverit animae apparere grati saporis, ipsi aliquando insipidi 〚v〛 vel amar〚{…}〛i videri poterit: cuius ratio est quod saliva a stomacho proveniens, atque humoris, quo abundat, qualitates semper retinens³¹⁵ sese immiscet exiguis particulis ciborum qui in ore sunt [41] eorumque 〚{…}〛actioni multum contribuit.

Odoratus^lxxxiv

³¹⁶Dependet quoque odoratus a plurimis exiguis filamentis,³¹⁷ quae a base cerebri versus nasum³¹⁸,lxxxv pendentes nullatenus aliter differunt a nervis tactui aut gustui inservientibus quam quod non egrediuntur capitis concavum quo totum cerebrum continetur: particulis³¹⁹ autem multo minoribus quam linguae nervi moveri possunt tum quia paulo sunt tenuiora, tum etiam quia magis immediate ab obiectis afficiuntur. 〚ipsa moventibus {…}〛³²⁰,lxxxvi ³²¹Cum enim haec machina [*superscriptum:* NB]³²²,lxxxvii respirat, aeris partes subtiliores per nares ingredientes penetrant per poros ossis 〚spongiosi〛 spongiosi,³²³ si non usque in concavitates cerebri ad minimum usque ad illud spatium quod est inter duas membranas quibus cerebrum involvitur, inde statim per pallatum via illis patet³²⁴ ut reciproce, cum aer pertectus

---

³¹² nullos poros habent, per quos linguae particulae, vel ad minimum salivae, quibus humectatur ingredi possunt, quemadmodum enim non poterunt efficere ut anima ullum sentiat saporem,] poris destituuntur, illae ipsae, sicuti efficere non poterunt, ut anima ullum sentiat saporem, S1; n'ont point aussi de pores en elles-mêmes, où les petites parties de la langue, ou bien pour le moins celles de la salive dont elle est humectée, puissent entrer; comme elles ne pourront faire sentir à l'âme aucun aucun goût, ni saveur, C1.

³¹³ § C1.

³¹⁴ mutatur temperamentum etiam 〚mutetur vis saporis〛 [*in margine:* gustus diminuetur]] mutatur temperamentum, efficacia itidem gustus minuatur S1; le tempérament de l'estomac se change, la force du goût se change aussi C1.

³¹⁵ semper retinens] imbuta S1; retient toujours C1.

³¹⁶ § 15 S1; § 34 C1.

³¹⁷ exiguis filamentis,] filamentis S1; petits filets, C1.

³¹⁸ nasum] nasum S1; le nez, au-dessous de ces deux petites parties toutes creuses, que les anatomistes ont comparées aux bouts des mamelles d'une femme, C1.

³¹⁹ particulis] particulis S1, 40; parties terrestres C1, 34.

³²⁰ magis immediate ab obiectis afficiuntur. 〚ipsa moventibus {…}〛]] ab obiectis magis immediate afficiuntur. S1; sont plus immédiatement touchés par les objets qui les meuvent. C1.

³²¹ § C1.

³²² [*Superscriptum:* NB] Cum enim haec] Cum enim haec S1; Car vous devez savoir que lorsque cette C1.

³²³ spongiosi,] cribriformis S1; qu'on nomme spongieux, C1.

³²⁴ inde statim per pallatum via illis patet] Unde illis via patet, qua se eodem tempore per palatum recipiant S1; d'où elles peuvent ressortir en même temps par le palais C1.

egreditur,³²⁵,ˡˣˣˣᵛⁱⁱⁱ [*in margine: //*]ˡˣˣˣⁱˣ spatium hoc per pallatum ingreditur regrediturque per nares³²⁶ atque in huius spacii orificio offendunt extremitates dictorum filorum, quae omnino [42] nuda sunt aut solum cooperta membranula admodum tenuia,ᶜᶜˣˡⁱᵛ quo fit ut non multum virium opus sit³²⁷ ut moveantur, ³²⁸quin etiam hi pori sic sunt constituti³²⁹ et adeo angusti, ut nullis particulis terrestribus ad haec filamenta transitum praebeant, quae sint crassioribus, quas³³⁰ propterea odores³³¹ superiusˣᶜ nominavi: nisi forsitan ex iis aliquae sunt quae³³² aquas vitae componunt, eo quod ipsarum figurae³³³ efficiant ut valde sint penetrantes, ³³⁴denique inter particulas hasce terrestres admodum exiguas quae paulo³³⁵ plus vel minus sunt crassiores aliis, aut quae propter ipsarum figuram magis aut minus sunt mobiles, animae praebere poterunt occasionem, ut diversos percipiat odores, imo illae tantum in quibus hi excessus valde sunt moderati, atque ab invicem admodum temperati, gratos ipsi ciebunt odores, nam quae ordinario tantum agunt nullo modo percipi poterunt, et quae magno aut exiguo nimium impetu [43] concitantur non poterunt illi aliter quam displicere.

Auditusˣᶜⁱ

³³⁶Quod ad exigua filamenta, quae organa sunt auditus,³³⁷ non opus est ut tam tenuia sint, quam praecedentia, sed sufficit modo sic sunt disposita, in aurium concavitatum fundo, ut facile omnia simul e⟦{...}⟧t e⟦{...}⟧odem modo, motibus illis tremulis, impellantur, quibus aer externus movet pelliculam tenuissimam praetensam orificio harum ⟦con⟧ concavitatum, ac nullo alio obiecto affici queant quam aer qui huic pelliculae subest.³³⁸ Hi enim tremuli ⟦{ip}⟧ impulsus ad cerebrum usque,

---

³²⁵ reciproce, cum aer pertectus egreditur,] reciproco quodam fluxu, cum aer exspirando e pectore expellitur, S1; réciproquement quand l'air sort de la poitrine; C1.

³²⁶ per nares] iterumque per nares S1; par le nez C1.

³²⁷ non multum virium opus sit] facile S1; n'ont pas besoin de beaucoup de force C1.

³²⁸ § C1.

³²⁹ quin etiam hi pori sic sunt constituti] Quin etiam illa est pororem constitutio S1; Quinetiam illa est pororum constitutio S2; Vous devez aussi savoir que ces pores sont tellement disposés, C1.

³³⁰ quas] iis, quas S1; que celles que C1.

³³¹ propterea odores] odores S1; odeurs pour ce sujet C1.

³³² aliquae sunt quae] quibusdam, quae eius sunt naturae, cuius illae sunt, ex quibus S1; quelquesunes de celles qui C1.

³³³ figurae] figuris, quas nactae sunt, S1; figure C1.

³³⁴ § C1.

³³⁵ quae paulo] quae semper uberiori copia aeri insunt, quam aliis corporibus mixtis, solae paulo S1, 41; qui se trouvent toujours en plus grande abondance dans l'air, qu'en aucun des autres corps composés, il n'y a que celles qui sont un peu C1.

³³⁶ § 16 S1; § 35 C1, 35.

³³⁷ auditus,] auditus S1; au sens de l'ouïe, C1.

³³⁸ subest.] incumbit, S1; est au-dessous C1.

nervorum ope, pertingentes, praebebunt animae occasionem sonorum ideam concipiendi. ⟦Et⟧ [339]Nota[340,xcii] unicum ex illis murmur[341] tantum aliquod excitaturum,[342] quod momento perit, et in quo nulla reperi[*subscriptum:* ∧; *superscriptum:* e]tur varietas, quam quatenus maius aut minus fuerit, prout magis vel minus impelletur auris. Verum cum plures[343] se invicem subsequuntur (ut ad oculum pa[44]tet in funium et campanorum sonantium tremoribus) fiet ut componant[344] sonum quem anima iudicabit dulciorem vel asperiorem prout erunt magis minusve aequales, acutiorem[345] vel graviorem quo promptius aut tardius se invicem subsequentur: adeo ut si vel dimidia, ⟦{t}⟧ vel tertia, vel quarta, vel quinta parte etc. promptius una vice quam alia se invicem sequantur, component sonum, quem iudicabit anima <u>acutiorem octava, aut quinta, aut quarta, aut tertia maiori etc.</u>[346,xciii] Et denique plures simul[347] toni convenient aut disconvenient, prout maior aut minor fuerit similitudo,[348] et intervalla inter exiguos impulsus[349] magis minusve aequalia [350]ut ex. gr. si divisiones linearum A B C D E F G [*in margine:* H][351] repraesentet exiguos impulsus, qui totidem varios sonos componunt, facili negotio iudicabimus illos qui repraesentantur lineis G et H non debere esse tam gratos auribus quam alios: [45] quemadmodum partes asperae[352] lapidis tactui tam gratae non sunt ac partes laeves[353] speculi perpoliti. Et oportet cogites B repraesentare sonum octava acutiorem quam A, C quinta, D quarta, E tertia maiori, et F tono[354,xciv] etiam maiori. Et nota A et B iuncta,[355] aut ABC aut

---

[339] § 36 C1.

[340] ⟦Et⟧ Nota] equidem, quod imprimis observandum S1, 41–42; Et notez C1.

[341] murmur] murmur inconditum S1, 42; bruit sourd C1.

[342] excitaturum,] auditui exhibebit, S1; pourra faire ouïr C1.

[343] plures] plures impulsus S1; plusieurs C1.

[344] componant] component S1; ces petites secousses composeront C1.

[345] acutiorem] acutior S1; et qu'elle jugera plus aigu C1, 36.

[346] <u>acutiorem octava, aut quinta, aut quarta, aut tertia maiori etc.</u>] acutiorem octava, aut quinta, aut quarta, aut tertia maiori, etc. S1; plus aigu d'une octave, ou d'une quinte, ou d'une quarte, ou d'une tierce majeure etc. C1.

[347] simul] simul S1; mêlés ensemble C1.

[348] similitudo,] similitudo, sive proportio: S1; rapport, C1.

[349] impulsus] impulsuum S1; secousses qui les composent C1.

[350] § C1.

[351] A B C D E F G [*in margine:* H]] A, B, C, D, E, F, G, H S1; A, B, C, D, E, F, G, H C1.

[352] partes asperae] asperitas S1, 43; parties raboteuses C1, 37.

[353] partes laeves] laevitas S1; celle C1.

[354] tono] sonum S1; ton C1.

[355] B iuncta,] B, S1; B joints ensemble, C1.

ABD aut etiam ABCC[356] magis convenire scilicet maiorem harmoniam effectura,[357] quam A, et F vel ACD[358] vel ADE etc.

[359]Quod mihi videtur sufficere, ut ostendam quo pacto animae, quae erit in machina quam describo, placere poterit musica quae observabit easdem regulas, quas nostra, imo quomodo ipsam multo perfectiorem efficere poterit, adminimum[xcv] si observemus suaviora non absolute sensibus esse gratiora, sed ea quae modo magis temperato sensus titillant: sicut sal et acetum saepe [[{...}]]linguae sunt gratiora, quam aqua dulcis[[,]]. [[e]]Et haec est ratio quare musica utatur tertiis, sextis, imo etiam aliquando dissonis, aeque ac unisonis, octavis et quintis.

[46] Visus[xcvi]

[360]Restat visus paulo exactius[361] explicandus, quia magis ad nostrum argumentum facit. Hic quoque sensus in hac machina dependet a duobus nervis compositis proculdubio ingenti numero parvorum capillamentorum maxime mobilium[362] in eum finem ut nimirum ad cerebrum referant diversas illas actiones particularum secundi elementi,[xcvii] quae iuxta id quod dictum supra praebebunt occasionem animae, cum huic machinae unita erit, varias colorum et luminis ideas concipiendi. [363]Verum quia oculi structura huic rei inservit[364] paucis describenda,[365] omissis de industria superfluis illis [[muni]] minutiis, quas ibi anatomicorum curiositas observat. [366]ABC est membrana satis crassa et dura, componens quoddam veluti vas,[367] receptaculum omnium aliarum oculi partium. DEF est membranula tenuior, intra priorem aulaei instar expansa. GHI est nervus vulgo opticus dictus,[368] cuius parva capillamenta HI[369] HI [47] per totum spatium ab H usque ad I[370,xcviii] diffusa totum oculi fundum tegunt. K L M tres sunt liquores valde pellucidi[371] totas has tunicas distendentes,

---

[356] ABCC] A B C E S1; ABCE C1.

[357] magis convenire scilicet maiorem harmoniam effectura,] maiori harmonia convenire, S1; sont beaucoup plus accordants C1.

[358] ACD] A G D S1; ACD C1.

[359] § 17 S1.

[360] § 18 S1; § 37 C1.

[361] exactius] exactius S1; plus exactement que les autres C1.

[362] maxime mobilium] summe mobilium S1, 44; les plus déliés, et les plus aisés à mouvoir qui puissent être C1.

[363] § 37 C1.

[364] inservit] perutilis et necessaria est S1; aide aussi C1, 38.

[365] paucis describenda,] paucis delineanda; S1; il est ici besoin que je la décrive; et pour plus grande facilité je tâcherai de le faire en peu de mots, C1.

[366] § 38 C1.

[367] vas,] rotundum vas, S1; un vase rond, C1.

[368] nervus vulgo opticus dictus,] nervus, vulgo opticus dictus, S1; le nerf, C1.

[369] HI] H, I, S1; HG, C1.

[370] per totum spatium ab H usque ad I] per totum spatium A, B, H usque ad I S1; tout autour, depuis H jusques à G et I C1.

[371] pellucidi] pellucidi S1, 45; claires et transparentes C1, 39.

figura qua s[ingu]los hic delineatos videtis.³⁷²In priori membrana pars BCB pellucida est, et magis gibba quam residuum: et refractio radiorum ingredientium fit versus perpendiculum. In altera³⁷³ superficies interior partis EF fundum oculi respiciens, tota obscura et nigra est, habetque in medio anterioris partis³⁷⁴ rotundum foramen exiguum, foris respi⟦s⟧cientibus nigerrimum apparens, quod pupillam appellamus. Non autem semper eadem magnitudine patet hic hiatus: sed EF pars secundae membranulae in qua est,³⁷⁵ liberrime innatans liquidissimo humori K, speciem exigui musculi habet, qui deducitur aut contrahitur, dirigente cerebro, pro ut usus petierit. ³⁷⁶Figura humoris L,³⁷⁷ qui crystallinus dicitur, similis est vitrorum, quae secundo libro³⁷⁸ descripsimus, quorum ⟦omnes⟧ [48] ope omnes radii ex uno puncto venientes in aliud colliguntur: et constat ex materia minus molli et firmiori, et per consequens maiorem causatur refractionem quam duo reliqui humores ipsum cingentes. ³⁷⁹E N sunt plurima filamenta nigra orta³⁸⁰ ex membrana DEF inde ubi tertia terminatur,³⁸¹ undiquaque humorem crystallinum amplexa, quae speciem perexiguorum tendinum prae se ferunt, et eorum ope hic humor pro intentione, {qua} visus noster in res propinquas [*in margine: //*]ˣᶜⁱˣ aut longe dissitas fertur, mox in maiorem gibbum curvatus, mox ⟦p⟧ magis in planum porrectus, totam oculi figuram nonnihil immutat.³⁸² Denique O O sunt 6 aut septem musculi extrinsecus oculo affixi, quorum ope promptissime et facillime³⁸³ quaquaversum moveri potest. ³⁸⁴Membrana autem BCB et tres humores K L M cum valde sint pellucidi,³⁸⁵ nullo modo impediunt quo minus radii per pupillae foramen³⁸⁶ ingredientes ad fundum [35]ᶜ oculi, ubi nervus est, et non minus facile in ipsum agant, quam si prorsus denudatus foret:³⁸⁷ imo ei conservando plurium inserviunt contra aeris iniurias, aliorumque corporum

---

³⁷² § C1.

³⁷³ In altera] Secundae membranae S1; en la deuxième peau C1.

³⁷⁴ in medio anterioris partis] in medio anterioris partis S1; au milieu C1.

³⁷⁵ membranulae in qua est,] membranulae S1; la peau dans laquelle il est, C1.

³⁷⁶ § C1.

³⁷⁷ L,] L S1; marquée L, C1.

³⁷⁸ secundo libro] secundo (Dioptrices) libro S1; au traité de la Dioptrique, C1.

³⁷⁹ § C1.

³⁸⁰ orta] introrsum producta S1; viennent du dedans C1.

³⁸¹ DEF inde ubi tertia terminatur,] D E F S1; D, E, F, C1.

³⁸² hic humor pro intentione, qua visus noster in res propinquas aut longe dissitas fertur, mox in maiorem gibbum curvatus, mox ⟦p⟧ magis in planum porrectus, totam oculi figuram nonnihil immutat.] crystallinus humor, quatenus visus noster in res propinquas, aut longe dissitas fertur, statim convexior redditur, aut planior, adeoque totam oculi figuram nonnihil immutat. S1, 45–46; sa figure se peut changer, et se rendre un peu plus plate, ou plus voûtée, selon qu'il est de besoin. C1, 39–40.

³⁸³ promptissime et facillime] facillime S1, 46; très facilement et très promptement C1, 40.

³⁸⁴ § 39 C1.

³⁸⁵ pellucidi,] pellucidi S1; claires et transparentes, C1.

³⁸⁶ per pupillae foramen] per pupillam S1; par le trou de la prunelle C1.

³⁸⁷ denudatus foret:] nudus esset, nullisque istiusmodi tunicis tegeretur. S1; était (…) découvert; C1.

externorum, a quibus tactu facili laedi posset; et ut ita mollis et tenellus[388] maneat, ut mirum non sit eum moveri posse ab actionibus tam parum sensi⟦{...}⟧bilibus, quales colores suppono.[389,ci]

[36][cii]

[390][49] quae[391] est in parte primae pellis notata BCB, et refractio quae in ea agitur est in causa cur radii, qui sunt circa partes oculi, venientes ad obiecta[392] intrare possint per pupillam et sic absque eo quod oculus moveatur, anima poterit videre maiorem numerum obiectorum, quam alias videre posset,[393] nam ex. gr. si radius PBKQ non incurvaret se ad punctum B non posset transire per puncta F F ut illa ratione ad nervum usque pervenire posset,[394] [395]nam refractio[396] quae fit in humore crystallino inservit ut visio magis fortis et simul magis distincta reddatur: sciendum enim est figuram talis humoris esse ita compassam[397,ciii] *[in margine: //][civ]* habita sc. ratione refractionum quae fiunt in aliis partibus oculi et simul etiam distantiae obiectorum ut quando visus versus aliquod punctum determinatum est directus, facit ut omnes radii qui ex illo puncto[398] in fundo oculi congreguntur[cv] praecise contra unam partem nervi qui in eo est,[399] et eadem ratione impedit ne aliqui aliorum radiorum, qui intrant [50] oculum tangant eandem partem istius nervii. [400]Nam ex. gr. cum humor crystallinus sit[401,cvi] dispositus ut respiciat punctum R facit ut omnes radii[402] RNS, RLS, et A[403,cvii] praecise in puncto S congregantur, eadem ratione

---

[388] tenellus] accuratum S1; délicat, C1.

[389] quales colores suppono.] quemadmodum sunt illae, quas nos colores appellamus, S1; comme sont celles que je prends ici pour les couleurs. C1, 41.

[390] § 19 S1; § 40 C1.

[391] quae] Convexitas, qua S1; La courbure qui C1.

[392] qui sunt circa partes oculi, venientes ad obiecta] venientes ab obiectis respicientibus eam oculi partem S1; qui viennent des objets qui sont vers les côtés de l'œil C1.

[393] quam alias videre posset,] quam si oculus ea parte non foret convexus. S1; qu'elle ne pourrait faire sans cela: C1.

[394] ut illa ratione ad nervum usque pervenire posset,] per ea ad nervum opticum. S1, 47; pour parvenir jusques au nerf. C1.

[395] § 41 C1.

[396] nam refractio] Refractio vero S1; La réfraction C1.

[397] compassam] compositam S1; compassée C1.

[398] ex illo puncto] ex illo puncto proveniunt S1; viennent de ce point, et qui entrent dans l'œil par le trou de la prunelle C1.

[399] praecise contra unam partem nervi qui in eo est,] accurate ad quoddam tunicae retinae sive nervi optici punctum. S1, 48; en un autre point au fond de l'œil, justement contre l'une des parties du nerf qui y est, C1.

[400] § C1, 42.

[401] humor crystallinus sit] humore crystallino sic S1; l'œil étant C1.

[402] facit ut omnes radii] radii S1; elle fait que tous les rayons C1.

[403] et A] etc. S1; etc. C1.

impedit quo minus aliqui ex illis[404] qui veniunt a punctis T, X et A[405,cviii] ad illud non perveniant, nam congregat [*in margine: //*][cix] etiam omnes puncti T circa punctum V et omnes puncti X circa punctum Y, et sic de caeteris: ubi alias si in illo loco[406] non fieret aliqua refractio, obiectum R tantum mitteret unum ex suis radiis ad punctum S et alii hunc per totum spatium VY dispergerentur, et pari ratione puncta T et X et omnia alia intermedia mitterent unum ex suis radiis versus idem punctum S, [407]evidens igitur est obiectum R fortius [*subscriptum:* ∧][cx] [*in margine: //*][cxi] posse agere[408] contra partem nervi qui est in illo puncto S quando ad illud mittit magnum numerum radiorum quam si O[409] tantum unicum mit[51]teret, et illam partem nervi S debere distinctius et certius cerebro referre actionem istius obiecti R cum ab illo solo radios recipiat, quam si alias[cxii] ex diversis aliis reciperet.[410] [411]Color niger tam ex superficie concava pellis EF quam parvorum filorum E N inservit etiam ad reddendum visionem magis distinctam, nam secundum ea quae dicta sunt su⟦per⟧pra de natura talis coloris illa diminuit vires radiorum, qui a fundo oculi versus anteriorem partem reflectuntur et etiam impedit[412] ne inde iterum redeant ad fundum oculi, ubi possent adferre aliquam confusionem. Si enim ex. gr. radii obiecti X darent[413] ad punctum R[414] contra nervum qui est albus reflectunt illum[415] hinc ex omnibus partibus versus N et versus F, unde iterum possent se reflectere versus S et versus V et T perturbare[416] si corpora N et F non forent nigra. [417]Mutatio autem[418] quae fit in humore crystallino inservit ad hoc, ut obiecta quae sunt in diversis spatiis[419] eorum [52] imagines in fundo oculi distincte possint depingere, nam secundum id quod dictum est in libro 2.°[420] si ex. gr. humor LN sit talis figura ut possit facere quod omnes radii puncti[421]

---

[404] eadem ratione impedit quo minus aliqui ex illis] Similiter ista humoris crystallini dispositio prohibet, ne radii S1; et empêche par même moyen, qu'aucun de ceux C1.

[405] et A] etc. S1; etc. C1.

[406] in illo loco] oculo S1, 49; dans cet œil C1.

[407] § C1.

[408] fortius [*subscriptum:* ∧] posse agere] fortius agere posse S1; doit agir plus fort C1.

[409] si O] si S1; s'il C1.

[410] quam si alias ex diversis aliis reciperet.] quam si a pluribus. S1; que si elle en recevait de divers autres. C1.

[411] § 42 C1.

[412] et etiam impedit] obtenebratque, et impedit S1; et empêche C1, 43.

[413] darent] tendentes S1; donnant C1.

[414] R] Y S1; Y C1.

[415] reflectunt illum] reflectuntur S1; se réfléchissent C1.

[416] V et T perturbare] V; atque ibidem actionem R et T turbare S1; V, et y troubler l'action des points R et T C1.

[417] § 43 C1.

[418] Mutatio autem] Mutatio vero S1, 50; Le changement de figure C1.

[419] quae sunt in diversis spatiis] remotiorum, aut propinquiorum S1; qui sont à diverses distances C1.

[420] libro 2.°] libro (Dioptrices) secundo S1; traité de la Dioptrique C1.

[421] puncti] puncti S1; qui partent du point C1.

R vadant[422,cxiii] praecise tangere [*in margine:* //][cxiv] nervum puncti[423] S eadem[424] absque eo quod mutetur non poterit facere quod radii puncti T quod est propinquius, aut puncti X quod est remotius, etiam ad illud perveniant, sed faciet ut radius TL eat[425] versus K[426] et TN versus G et e contra quod XL ibit[427] versus G, et XN versus K,[428] et sic de caeteris, ita ut repraesentare possit distincte punctum X necessarium est totam figuram istius humoris LN[429] mutari et fiat parum magis dimissior,[430] sicuti illa quae est notata I[431] ut repraesentet punctum T necessarium est ut fiat parum magis densa[432] sicuti illa quae est notata F[433] [434]mutatio quippe magnitudinis quae accidit [53] pupillae inservit ad moderandas vires visus, nam necesse est ut sit minor[435] quando lumen est nimis vehemens, ut in oculum tam multi radii non intrent ne illa ratione nervus offendatur, et ut sit maior quando lumen est nimis debile, ut sufficienter intrent ut percipi possit, et quidem posito quod lumen maneat aequale necesse est ut pupilla sit maior quando obiectum quod respicitur ab oculo est remotum, quam quando est propinquum. Nam ex. gr. si per pupillam oculi[436] non intrent plures radii puncti R quam sint necessarii ut possit percipi[437] necesse est ut tantundem intrent in oculum[438] et consequenter pupillam esse maiorem. [439]Parvitas autem pupillae est etiam ad hoc ut reddat visionem magis distinctam, sciri enim debet quod qualemcumque figuram habere possit humor crystallinus impossibile sit illam facere ut radii qui veniunt a diversis punctis obiecti congregentur omnes exacte et praecise in aeque multis aliis punctis. Sed si ex. gr. radii puncti R congregentur praecise in puncto S nulli erunt ex puncto T praeter illos, qui [54] transeunt per circumferentiam et per centrum unius circulorum quod potest describi in superficie istius humoris crystallini qui exacte possint congregari in puncto V et consequenter alii qui eo erunt minori numero quo pupilla erit minor, tangentes nervum in aliis punctis non poterit aliter fieri quin ibi adferant aliquam confusionem. Unde fit ut si

---

[422] possit facere quod (…) vadant] dirigere possit S1; fasse que (…) aillent C1.
[423] tangere nervum puncti] ad punctum S1; toucher le nerf au point C1, 44.
[424] eadem [figura]] eadem [figura] S1; la même humeur C1.
[425] eat] tendat S1; ira C1.
[426] K] K S1; H C1.
[427] ibit] tendat S1; ira C1.
[428] K,] K. S1; H, C1.
[429] LN] LN S1; NL C1.
[430] parum magis dimissior,] planiorem obtusioremve. S1, 51; un peu plus plate, C1.
[431] I] litera i S1; I C1.
[432] magis densa] convexior S1; plus voûtée C1.
[433] F] litera F S1; F C1.
[434] § 44 C1.
[435] necesse est ut sit minor] imminuitur S1; il est besoin qu'elle soit plus petite C1.
[436] oculi] oculi 7 S1; l'œil 7 C1, 45.
[437] possit percipi] visu possit percipi S1; pouvoir être sentis C1.
[438] oculum] oculum 8 S1; l'œil 8 C1.
[439] § 45 C1.

visio eadem⁴⁴⁰ eiusdem oculi una vice sit minus vehemens quam altera erit etiam minus distincta,⁴⁴¹ sive hoc proveniat ex distantia obiecti, sive ex tenuitate luminis.⁴⁴² ⁴⁴³Inde etiam sit^cxv q⟦{...}⟧uod anima nunquam distinctissime videre poterit singulis visionibus⁴⁴⁴ praeter unicum punctum, illud sc. circa quod omnes partes oculi pro tunc erunt directae,⁴⁴⁵ et sic alia ei apparebunt eo confusiora, quo magis hoc distabunt: nam si ex. gr. radii puncti R congregentur⁴⁴⁶ in puncto S radii puncti X congregabuntur adhuc minus exacte in puncto Y, quam [55] radii puncti T congregabuntur in puncto V et sic de aliis iudicandum est, proportione sc. servata inter illos qui magis distant⁴⁴⁷ a puncto R musculi autem O O levissime undequaque vertentes oculos ad hoc sunt ut vi⟦d⟧tetur⁴⁴⁸ ille defectus, nam minimo instanti temporis possunt omnibus punctis obiecti eum successive applicare et sic facere ut anima illa omnia unum post alterum videre possit exactissime.

⁴⁴⁹Non addo hic peculiariter illud occasione cuius anima⁴⁵⁰ illas differentias omnes colorum percipere poterit de iis enim satis diximus hic supra, neque etiam inquirimus⁴⁵¹ quaenam obiecta quae proponuntur visum delectent, quinam non delectent, nam ex iis quae de aliis sensibus diximus facile est intellectu quod lume⟦{...}⟧n nimis vehemens debeat offendere visum, moderatum autem illum recreare, et quod inter omnes colores⁴⁵² color viridis, [56] qui consistit proportione⁴⁵³ unius ad duo, sit quasi octava inter consonantes⁴⁵⁴ musicae, et panis inter cibos,⁴⁵⁵ hoc est ille qui magis universaliter delectat, et denique omnes illi diversi et varii istiusmodi colores,⁴⁵⁶,^cxvi qui saepe multo magis recreant quam color viridis,

---

⁴⁴⁰ visio eadem] visus S1, 54; la vision C1, 46.

⁴⁴¹ erit etiam minus distincta,] obiecta quoque minus distincte referat; S1; elle sera aussi moins distincte, C1.

⁴⁴² luminis.] luminis S1; de la lumière; parce que la prunelle étant plus grande quand elle est moins forte, cela rend aussi la vision plus confuse.

⁴⁴³ § 46 C1.

⁴⁴⁴ singulis visionibus] singulis visionibus S1; à chaque fois C1.

⁴⁴⁵ omnes partes oculi pro tunc erunt directae,] totus oculus convertitur. S1; toutes les parties de l'œil seront dressées pour lors, C1.

⁴⁴⁶ congregentur] omnes (...) exacte congregentur S1; s'assemblent tous exactement C1.

⁴⁴⁷ proportione sc. servata inter illos qui magis distant] proportione scilicet servata inter ea, quae magis distant S1; à mesure qu'ils sont plus éloignés C1.

⁴⁴⁸ ut vi⟦d⟧tetur supplent; S1; à suppléer C1.

⁴⁴⁹ § 20 S1; § 47 C1.

⁴⁵⁰ anima] animae S1; cette âme C1.

⁴⁵¹ inquirimus] prolixe describo S1, 55; dis C1.

⁴⁵² inter omnes colores] maxime S1; entre les couleurs C1, 47.

⁴⁵³ proportione] in proportione S1; en l'action la plus modérée (qu'on peut nommer par analogie la proportion C1.

⁴⁵⁴ consonantes] consonantes sonos S1; consonances C1.

⁴⁵⁵ panis inter cibos,] panis quotidianus ad caetera cibi genera, S1; le pain entre les viandes que l'on mange, C1.

⁴⁵⁶ varii istiusmodi colores,] colores, quibus novitas gratiam conciliat, S1; couleurs de la mode, C1.

sunt quasi concordantiae et transitus boni musici, aut ....^cxvii boni coqui,^457 qui multo magis afficiunt sensus, et instanti recipiunt multo maiorem delectationem,^458 sed etiam multo citius illis parent nauseam^459 quam obiecta simplicia et ordinaria facient.

^460Solum adhuc debemus dicere quid sit occasione^461 cuius anima poterit percipere situm, figuram, distantiam, magnitudinem et alias similes qualitates, quae referunt se ad unum^462 sensum particularem, sicuti faciunt illae de quibus hactenus fuimus locuti et visui^463 sint communes, imo etiam aliquomodo aliis sensibus.

[57] ^464Nota igitur quod manus A ex. gr. tangat corpus C partes cerebri B inde fit quod^465,cxviii parva fila illorum^466,cxix nervorum alio modo erunt disposita, quam si tangeret unum quod foret alterius figurae, aut alterius magnitudinis, aut in alio loco sita,^ccxlv et sic quod anima occasione illorum^467 cognoscere poterit situm illius corporis, suam figuram et magnitudinem, et simul omnes alias qualitates,^468 et eadem ratione si oculus D sit directus versus obiectum E anima poterit cognoscere situm istius obiecti, quia nervi istius oculi alio modo erunt disposit[[...]]i, quam si in aliam partem esset directus, et sic suam figuram poterit^469 quia radii puncti I^470 congregentur in puncto 2 contra nervum qui dicitur opticus, et radii puncti 3 in puncto 4 et sic de caeteris, del[in]e[a]bunt aliam quae ad suam exacte se referet, et quod illa ex. gr.^471 poterit cognoscere distantiam puncti I^472 quia dispositio humoris crystallini erit alterius [58] figurae ut omnes radii venientes ab hoc puncto praecise in fundo oculi simul congregentur^473 in puncto 2, quod tenere medium suppono, quam

---

^457 concordantiae et transitus boni musici, aut .... boni coqui,] egregiae, atque insolitae excellentis musici modulationes, aut sapientis coqui delicatior cibus. S1; les accords et les passages d'un air nouveau, touché par quelque excellent joueur de luth, ou les ragoûts d'un bon cuisinier, C1.

^458 multo magis afficiunt sensus, et instanti recipiunt multo maiorem delectationem,] in principio sensum velut prurigine suavius afficiunt, S1; chatouillent bien davantage le sens, et lui font sentir d'abord plus de plaisir, C1.

^459 parent nauseam] taedium, nauseamque pariunt, S1; le lassent C1.

^460 § 48 C1.

^461 occasione] rationem S1; moyen C1.

^462 ad unum] non uno S1; pas à un seul C1.

^463 visui] tactui, gustui S1; à l'attouchement et à la vue C1.

^464 § C1, 48.

^465 inde fit quod] unde (...) proveniunt S1, 57; d'où viennent C1.

^466 illorum] illorum S1; ses C1.

^467 occasione illorum] per peculiarem illum modum, quo afficitur, S1; par leur moyen C1.

^468 et simul omnes alias qualitates,] similesque qualitates. S1; et toutes les autres semblables qualités. C1.

^469 poterit] agnoscet S1, 58; pourra connaître C1, 49.

^470 I] 1 S1; 1 C1.

^471 illa ex. gr.] Eadem S1; elle (...) par exemple C1.

^472 I] 1 S1; 1 C1.

^473 simul congregentur] colligantur S1; s'assemblent C1.

si esset magis propinquum aut magis remotum⁴⁷⁴ ut supra dictum est, et ut ulterius illa poterit cognoscere figuram puncti 3 et simul omnium aliorum, inde⁴⁷⁵,ᶜˣˣ radii eodem tempore intrabunt oculum, quia humor crystallinus, cum sic sit dispositus radii istius puncti 3 non adeo praecise in puncto 4 congregabuntur, quam puncti {[[I]]1}⁴⁷⁶ in puncto 2 et sic de caeteris, et quod proportione servata eorum actio non erit omnino aeque vehemens⁴⁷⁷ sicut et ante⁴⁷⁸ dictum est. Denique anima poterit cognoscere magnitudinem obiectorum visus et omnes alias⁴⁷⁹ similes qualitates sola notitia quam habebit distantiae et situs punctorum eorum⁴⁸⁰,ᶜˣˣⁱ sicuti et reciproce de eorum distantia iudicabit per solam [59] opinionem quam habebit de eorum⁴⁸¹,ᶜˣˣⁱⁱ magnitudine.

⁴⁸²Nota et[iam] quod utraque manus F G⁴⁸³ teneat unaqueque baculum HI⁴⁸⁴ quibus tangat obiectum K etiamsi anima ignoret aliunde longitudinem istorum baculorum, tamen quia novit distantiam quae est inter duo⁴⁸⁵ puncta I⁴⁸⁶ et magnitudinem oculorum,⁴⁸⁷ illa tanquam geometria naturalis poterit cognoscere ubinam sit obiectum K et eadem radioneᶜˣˣⁱⁱⁱ si ambo oculi L et M sint directi versus obiectum N magnitudo lineae LM et simul angulorum LMN, MML⁴⁸⁸ dabunt illi cognoscere ubinam sit punctum N, ⁴⁸⁹sed etiam in his omnibus satis saepe poterit decipi, nam primo si situs digiti aut oculi⁴⁹⁰ aliqua causa exteriore sit contrarius,⁴⁹¹ illa non tam facile concordabit cum eo quod est in minimis partibus⁴⁹² cerebri unde profluunt nervi, quasi vero dependeret tantum a musculis,⁴⁹³ et sic anima quae tantum mediantibus illis partibus

---

⁴⁷⁴ tenere medium suppono, quam si esset magis propinquum aut magis remotum] medium inter remotum et propinquum esse suppono S1; je suppose en être le milieu, que s'il en était plus proche ou plus éloigné, C1.

⁴⁷⁵ inde] ex quibus S1; dont C1.

⁴⁷⁶ puncti {[[I]]1}] illi, qui veniunt a puncto 1 S1; ceux du point 1 C1.

⁴⁷⁷ eorum actio non erit omnino aeque vehemens] tantis viribus agent S1; leur action ne sera pas du tout si forte C1.

⁴⁷⁸ ante] alias S1; tantôt C1.

⁴⁷⁹ visus et omnes alias] visum afficientium, aliasque S1; de la vue, et toutes leurs autres C1.

⁴⁸⁰ punctorum eorum] punctorum omnium, quibus [anima] constat S1; de tous leurs points C1, 50.

⁴⁸¹ eorum] eius [*i.e.:* animae] S1; leur C1.

⁴⁸² § C1.

⁴⁸³ F G] F et G S1, 59; f et g C1.

⁴⁸⁴ HI] H et I S1; i et h C1.

⁴⁸⁵ quae est inter duo] duorum S1; qui est entre les deux C1.

⁴⁸⁶ I] F et G S1; f et g C1.

⁴⁸⁷ oculorum,] angulorum F G H, et G F I, S1; des angles fgh, et gfi, C1.

⁴⁸⁸ MML] M L N S1; MLN C1.

⁴⁸⁹ § 49 C1.

⁴⁹⁰ digiti aut oculi] digiti oculive S1, 60; de la main, ou de l'œil, ou du doigt C1, 51.

⁴⁹¹ sit contrarius,] contra naturam distorqueatur, S1; est contrainte C1.

⁴⁹² cum eo quod est in minimis partibus] particulis S1; avec celle des petites parties C1.

⁴⁹³ musculis,] musculis naturaliter agentibus dependeret; S1; muscles; C1.

[60] cerebri illius sentiet, proculdubio illa decipietur, [494]prout hic verbi gr. si manus F sit ex se ipsa disposita ut vertat se versus O et interea temporis ⟦cogitatur⟧ cogatur,[495] aliqua vi exteriore, vertere se versus K partes cerebri, unde illi[496,cxxiv] nervi profluunt non erunt omino eodem modo[497] dispositae quam si vi musculorum[498] essent ita directae[499] versus A[500] nec etiam eodem modo si vere directae essent[501] versus O sed modo media[502,cxxv] inter illa duo si[503,cxxvi] eodem modo quo si essent directae[504] versus V[505] et sic occasione illorum[506] anima iudicabit quod obiectum K sit in puncto P et quod sit aliud quam illud quod tactum est per eius manum G [507]eodem modo si oculus[508]vi obiect⟦{…}⟧i N[509,cxxvii] a me sit[510] retractus [in margine: //][cxxviii] et ita dispositus quasi deberet respicere versus Q anima tunc iudicabit illum esse directum versus R, et quia in hoc situ[511] radii obiecti N intra⟦eodem⟧bunt intra,[512] eodem modo quo intrarent radii puncti S [61] ⟦v⟧ si[513] vere esset directus versus R iudicabit enim illud obiectum[514] sit in puncto S et quod sit aliud quam illud quod respicitur ab oculo M[515] [516]eodem modo etiam ambo digiti T V[517] tangentes parvum globulum X dabunt[518] animae iudicare quod tangant duo diversa[519] quia essent per modum crucis

---

[494] § C1.

[495] interea temporis ⟦cogitatur⟧ cogatur,] distorqueatur S1, 61; contrainte C1.

[496] illi] hi S1; ses C1.

[497] omino eodem modo] eo modo S1; tout à fait (…) en même sorte, C1.

[498] vi musculorum] spontaneo musculorum motu S1; par la force de ses muscles C1.

[499] partes cerebri (…) essent ita directae] manus (…) esset conversa S1; la main fût ainsi tournée C1.

[500] A] K S1; K C1.

[501] [partes cerebri] directae essent] [manus] esset directa S1; elle [main] était (…) tournée C1.

[502] media] media [errata corrige: medio] S1; d'une façon moyenne C1.

[503] si] scilicet S1; savoir C1.

[504] [partes cerebri] essent directae] [manus] esset conversa S1; elle [main] était tournée C1.

[505] V] P S1; P C1.

[506] occasione illorum] occasionem capiet S1; la disposition que cette contrainte donnera aux parties du cerveau fera C1.

[507] § C1, 52.

[508] oculus] oculus M S1; l'œil M C1.

[509] N] N S1; N C1.

[510] a me sit] sit S1; est C1.

[511] in hoc situ] oculo sic constituto S1, 62; en cette situation C1.

[512] intra⟦eodem⟧bunt intra,] in oculum incident plane, S1; entreront dans l'œil, C1.

[513] intrarent radii puncti S ⟦v⟧ si] illi oculos feriunt, qui veniunt a puncto S, si oculus S1; que feraient ceux du point S, si l'œil C1.

[514] obiectum] obiectum N S1; objet N C1.

[515] oculo M] oculo M S1; l'autre œil C1.

[516] § C1.

[517] T V] T et R S1; t et v C1.

[518] dabunt] occasionem dabunt S1; feront C1.

[519] diversa] corpora diversa S1; différentes C1.

posita et vi retentae extra eorum situm naturalem, [520]et quod magis est si radii aut aliae lin⟦i⟧eae vi, quorum actiones obiectorum distantiam transeunt per sensus,[521],[cxxix] sint incurvatae, anima quae communiter supponet illas esse rectas poterit facillime decipi. Ut si baculus H^{522} sit incurvatus versus K videbitur animae quod obiectum K quod tangit[523] sit versus R^{524} et si oculus L radios obiecti N per vitrum 3^{525} recipiat, quod illud incurvat[526] videbitur animae quod illud obiectum sit versus A^{527} et eadem ratione si oculus B recipiat radios puncti D per vitrum C quos omnes eodem modo ⟦s⟧ curvatos[528] suppono, quam si ex puncto E venirent [62] et quam si illi qui sunt in puncto F ex puncto G venirent et sic de caeteris, et tunc animae[529] videbitur quod obiectum DFH sit aeque distans et aeque magnum ac apparet obiectum EGI [530]et ut concludamus notandum est omnes modos, quibus anima poterit cognoscere distantiam obiectorum visus,[531] esse incertos:[532] nam quantum ad angulos LMN, MNL,[533] et eorum similes, nunquam fere magis sensibiliter mutantur, quando obiectum distat magis quam 15 aut 10^{534} pedibus, et quod attinet ad dispositionem humoris crystallini illa adhuc multo minus[535] sensibiliter mutatur statum[536] atque obiectum distat plus quam tribus aut quatuor pedibus ab oculo. Et ⟦q⟧ denique quantum ad hoc quod est iudicare de distantiis sola opinione magnitudinis obiectorum vel quia radii qui a diversis eorum punctis veniunt non congregantur aeque exacte in fundo oculi,[537] exemplum picturarum perspectivarum satis ostendit[538] quam sit facile decipi: [63] nam quando earum figurae sunt minores quam nobis imaginemur illas debere esse,

---

[520] § 50 C1.

[521] lin⟦i⟧eae vi, quorum actiones obiectorum distantiam transeunt per sensus,] lineae, interveniente quavis obiectorum distantium actione, in sensus illabentes S1, 62–63; lignes, par l'entremise desquelles les actions des objets éloignés passent vers les sens, C1.

[522] H] H S1, 63; HY C1, 53.

[523] quod tangit] quod tanget S1; que ce bâton touche C1.

[524] R] R; Y C1.

[525] 3] 3; Z C1.

[526] illud incurvat] eos frangit sive incurvat S1; les courbe C1.

[527] versus A] in puncto A S1; vers A C1.

[528] curvatos] fractos sive incurvatos S1, 64; les plier C1.

[529] et tunc animae] animae S1; à l'âme C1.

[530] § 51 C1, 54.

[531] visus,] visu S1; de la vue, C1.

[532] incertos:] dubios (…) atque incertos. S1; incertains: C1.

[533] MNL,] M L N, S1; MLN, C1.

[534] 10] 20 S1; vingt C1.

[535] multo minus] minus S1; moins C1.

[536] mutatur statum] mutatur S1; change C1.

[537] oculi,] oculi S1; de l'œil les uns que les autres, C1, 55.

[538] exemplum picturarum perspectivarum satis ostendit] tabulae ad perspectivae amussim depictae, utriusque argumenti vim elidunt, ostenduntque S1; l'exemple des tableaux de perspective nous montre assez C1.

et earum colores esse parum obscuros, et lineamenta parum confusa, hoc fit quod videantur nobis magis distare,[539] et maiores quam revera sint.

[540]Postquam autem quinque sensus externos tales quales sunt in hac machina sic explicaverimus, nobis etiam aliquid dicendum est de aliquibus sensationibus[541] quae in ea reperiuntur.

[542]Quando liquores[543] quos hic supra inservire dixi, tanquam aqua fortis in stomacho ut continuo intre[n]t massa[544] sanguinea per extremitates arteriarum, non inveniunt satis multos cibos dissolvendos ut occupe[n]t omnes eorum vires, illam[cxxx] in stomachum ipsum vertunt, et parva fila nervorum eius fortius quam solito agitant, et sic moventur partes cerebri unde profluunt, qua ratione anima, unita isti machinae, ideam generalem famis concipiet, et si liquores isti sint potius nati[545] eorum actiones dirigere [64] ad certos cibos particulares quam ad alios, sicuti ordinari{e}[546,cxxxi] aqua fortis facilius saturat metalla quam seram,[cxxxii] et contra nervos[547] modo particulari agent, qua ratione anima pro tunc concipiet appetitum et desiderium[548] ad certos aliquos cibos potius quam ad alios. (Hic notari potest mira huius machinae conformatio, quod fames oriatur ex ieiunio; sanguis enim circulatione acrior fit, et ita liquor ex eo ad stomachum veniens nervos magis vellicat idque modo peculiari, si peculiari sit constitutio sanguinis, unde[549,cxxxiii] pica mulierum).

[550]Liquores[551] autem isti praecipue congregantur in fundo stomachi, et ibi causant sensationem famis [552]sed etiam continuo plures eorum partes versus guttur ascendunt et quando in satis magna copia non veniunt ut humectent, et poros per modum aquae impleant tunc eo tantum ascendunt per modum aeris aut fumi, et pro tunc agitantes contra nervos suos alio modo quam solito, causant in cerebro motum occasione cuius anima concipiet ideam sitis: [553]eadem ratione[554] quando sanguis[555] magis purus et

---

[539] magis distare,] remotiores S1, 65; de beaucoup plus éloignées C1.

[540] § 21 S1; § C1.

[541] aliquibus sensationibus] sensus internos S1; certains sentiments intérieurs C1.

[542] § 22 S1; § 52 and beginning of part 4, C1.

[543] liquores] liquor S1; les liqueurs C1.

[544] massa] ex (...) massa S1; de toute la masse C1.

[545] sint potius nati] eius sit temperamenti S1; sont disposées (...) plutôt C1.

[546] ordinari{e}] communis S1, 66; commune C1, 56.

[547] nervos] nervos stomachi S1; les nerfs de l'estomac C1.

[548] appetitum et desiderium] appetitum (...) comedendi. S1; l'appétit de manger C1.

[549] unde] unde S1; inde S2; unde C1.

[550] § 23 S1.

[551] Liquores] Liquor S1; liqueurs C1.

[552] § 53 C1.

[553] § 54 C1, 57.

[554] eadem ratione] Pari ratione S1, 67; Ainsi C1.

[555] sanguis] sanguis S1; le sang qui va dans le cœur C1.

magis [65] subtilis est, et ibi ignescit facilius solito,⁵⁵⁶ nervum pororum⁵⁵⁷,ᶜˣˣˣⁱᵛ qui in eo est, secundum ea quae re⟦{per}⟧quiruntur disponit, ut causet sensationem laetitiae,⁵⁵⁸ et secundum illa ut sensationem tristitiae causet, quando sanguis ille qualitates omnino contrarias⁵⁵⁹ habet. ⁵⁶⁰Et ex iis omnibus facile est intellectu quid in machina illa sit quod se omnibus aliis sensibus internis referat qui in nobis sunt, ita ut iam sit tempus quod incipiam explicare quomodo spiritus animales habeant suum cursum in concavitatibus et poris cerebri, et quae functiones ex iis dependeant.

⁵⁶¹Si tibi aliquando fuerit curiositas propinquius organa templorum nostrorum⁵⁶² videre, scis satis quomodo spiracula aerem in certis receptaculis⁵⁶³,ᶜˣˣˣᵛ trudant, quae ut videtur mihi, eo fine vocantur vent⟦{...}⟧icularii, et quando aer iste ex illis in canales intret modo in hos modo in alios pro diversis modis quibus⁵⁶⁴ organista digitos movet super claviorium,⁵⁶⁵ ⁵⁶⁶iam autem hic concipere potes quomodo cor et arteriae quae spiritus animales in concavitatibus cerebri nostrae machinae [66] trudunt, sint quasi spiracula⁵⁶⁷ quae aerem in suis canalibus⁵⁶⁸ trudunt, quod obiecta exteriora quae sunt nervi quos movent,⁵⁶⁹,ᶜˣˣˣᵛⁱ faciant ut spiritus istarum concavitatum intrent in aliquos ex istis poris,⁵⁷⁰,ᶜˣˣˣᵛⁱⁱ sunt quasi digiti organistae, qui secundum diversos modus,ᶜˣˣˣᵛⁱⁱⁱ quos⁵⁷¹ imprimunt, faciunt ut aer ex spiraculis⁵⁷² in aliquem canalem intret. Cum autem suavitas et dulcedo⁵⁷³ organorum non dependeat ex recta dispositione suorum canalium,⁵⁷⁴ qui exterius apparent, nec etiam ex figura spiraculorum⁵⁷⁵ aut aliquarum aliarum partium, sed solum ex tribus, sc. ex

---

⁵⁵⁶ et ibi ignescit facilius solito,] citiusque effervescit, S1; et s'y embrase plus facilement qu'à l'ordinaire, C1.

⁵⁵⁷ nervum pororum] nervulum S1; le petit nerf C1.

⁵⁵⁸ laetitiae,] voluptatis S1; de la joie; C1.

⁵⁵⁹ qualitates omnino contrarias] contraria temperie S1; des qualités toutes contraires. C1.

⁵⁶⁰ § C1.

⁵⁶¹ § 24 S1; § 55 C1.

⁵⁶² organa templorum nostrorum] pneumatica templorum organa S1; les orgues de nos églises C1.

⁵⁶³ receptaculis] venti receptacula S1; réceptacles C1.

⁵⁶⁴ pro diversis modis quibus] prout S1; selon les diverses façons que C1.

⁵⁶⁵ claviorium,] organi claves, sive manubria epistomiorum. S1; le clavier. C1.

⁵⁶⁶ § 25 S1.

⁵⁶⁷ spiracula] organi folles S1; les soufflets de ces orgues, C1.

⁵⁶⁸ suis canalibus] ipsa venti receptacula S1; les porte-vents C1.

⁵⁶⁹ quae sunt nervi quos movent,] quae, quatenus nervos movent, S1, 68; qui, selon les nerfs qu'ils remuent, C1.

⁵⁷⁰ poris,] nervorum poros, S1; pores C1.

⁵⁷¹ diversos modus, quos] diversos motus, quibus claves S1; les touches qu'ils C1, 58.

⁵⁷² spiraculis] venti receptaculis S1; porte-vents C1.

⁵⁷³ suavitas et dulcedo] harmonia S1; l'harmonie C1.

⁵⁷⁴ recta dispositione suorum canalium,] illa tuborum, quibus excitatur, dispositione, S1; cet arrangement de leurs tuyaux C1.

⁵⁷⁵ spiraculorum] receptaculorum venti S1; leurs porte-vents C1.

aere qui ex spiraculis[576] profluit, ex canalibus qui sonum reddunt, et ex distributione aeris in istos canales, sic igitur functiones,[577] de quibus hic,[578] nullo modo dependent ex exteriore figura omnium illarum partium visibilium, quas in substantia cerebri distinguunt astronomi,[cxxxix] nec etiam ex figura concavitatum, sed tantum ex corde[579] et poris [[cerb]] cerebri per quos transeunt, modo[580,cxl] quo distribuuntur illi spiritus [67] in istis poris. Itaque tantum est necesse ut per omnia explice[[{...}]]m, quod[581] in his tribus magis debeat considerari. [582]Primo[cxli] quantum ad spiritus animales attinet, illi in m[[in]]aiori aut minori copia, et eorum partes magis aut minus crassae, magis aut minus agitatae, magis aut minus aequales,[583] inter se mutuo esse possunt una vice, quam altera:[584] et ratione istarum differentiarum[585] fit ut omnes humores[586] diversi, et inclinationes[587] naturales quae in nobis sunt, saltem quatenus non dependent ex constitutione cerebri nec affectibus particularibus animae, in hac machina repraesentantur. Si enim spiritus isti in maiori copia sint quam soleant {sunt}[[{...}]] ad hoc nervi ut [[{ppii}]][cxlii] proprii ut in illa[588] motus omnino similes excitent illis qui in nobis exhibent bonitatem, liberalitatem, et amorem, et similes illis qui confidentia et audaciam exhibent,[589] si sc. eorum partes [[{sunt}]] sint fortes, et secundum figuram crassiores:[590,cxliii] promptitudinem et diligentiam, et desiderium si illae sint magis agitatae, et ingenii tranquillitatem si in earum [[{...}]] agitatione sint magis aequales, sicuti econtra malignitatem, timiditatem, inconstantiam, [68] negligentiam, et inquietudinem exhibent,[591] si qualitatibus illi careant. [592]Scias autem omnes alios humores,[593] aut naturales inclinationes[594] esse compositas, aut dependere ex

---

[576] spiraculis] follibus S1; soufflets C1.

[577] sic igitur functiones,] Similiter functiones, S1; ainsi je veux vous avertir, que les fonctions C1.

[578] hic,] hic sermo, S1; ici question, C1.

[579] ex corde] ex spiritibus, qui ex corde profluunt S1; des esprits qui viennent du cœur C1.

[580] modo] et a modo S1; et de la façon C1.

[581] per omnia explice[[{...}]]m, quod] explicare, quicquid S1; explique par ordre tout ce C1.

[582] § 26 S1, 69; § 56 C1.

[583] aequales,] inaequales. S1; égales C1.

[584] una vice, quam altera:] diversis temporibus S1; une fois que l'autre; C1.

[585] differentiarum] differentiarum S1; quatre différences C1.

[586] humores] complexiones S1; humeurs C1.

[587] inclinationes] affectus S1; inclinations C1.

[588] illa] illa machina S1; elle C1.

[589] exhibent,] indicant nos esse praeditos S1; témoignent en nous C1, 59.

[590] fortes, et secundum figuram crassiores:] fortiores, crassioresque. S1; plus fortes et plus grosses; et de la constance, si avec cela elles sont plus égales en figure, en force, et en grosseur; C1.

[591] exhibent,] prae se ferunt, S1; ces mêmes esprits sont propres à exciter en elle des mouvements tout semblables à ceux qui témoignent en nous C1.

[592] § C1.

[593] humores,] complexiones, S1; humeurs C1.

[594] inclinationes] cupiditates S1; inclinations C1.

his iam denotatis:⁵⁹⁵ sic humor⁵⁹⁶ laetus componitur ex promptitudine et ingenii tranquillitate, bonitas autem et confidentia ad hoc sunt, ut humorem⁵⁹⁷ illum magis perfectum reddant. Humor⁵⁹⁸ tristis componitur ex tarditate et inquietudine, potest autem augeri per malignitatem et timiditatem.⁵⁹⁹ Humor⁶⁰⁰ colericus componitur ex promptitudine et inquietudine, fortificatur⁶⁰¹ autem a malignitate et confidentia, et sic de caeteris. ⁶⁰²Quia vero humores⁶⁰³ iidem aut saltem passiones, ad quas disponunt multum, etiam dependent ex impressionibus quae in substantia ⟦{…}⟧cerebri fiunt, et quia ea iterum postea dicemus, sufficiet mihi hic dicere causas unde differentiae spirituum profluant.⁶⁰⁴

⁶⁰⁵Succus igitur ciborum qui ex stomacho per venas transit immiscens se cum sanguine, ei semper aliquas ex suis qualitatibus communicat, et inter alia eum ordinarie reddit magis [69] crassum tempore sc. quo fit commixtio,⁶⁰⁶ ita ut pro tunc partes minores istius sanguinis, quas sanguis mittit ad cerebrum ut ibi spiritus animales componant, non soleant ita esse agitatae, nec ita fortes nec ita abundantes, et per consequens non etiam reddere corpus istius machinae ita leve et alacre sicuti fit aliquo tempore post absolutam digestionem,⁶⁰⁷ quando sc. sanguis saepius transivit per cor, et sic subtilior fuit factus. ⁶⁰⁸Cum autem aer respirationis⁶⁰⁹ se etiam aliquando⁶¹⁰,ᶜˣˡⁱᵛ cum sanguine intermisceat, antequam intret in concavitatem sinistram cordis, facit ut se in eo fortius involvat,⁶¹¹,ᶜˣˡᵛ et spiritus magis vivos et magis agitatos producat tempore sicco quam humido, sicuti tunc experimur omne genus flammae magis ardere ⁶¹²quando hepar est bene dispositum, et sanguinem, qui debet intrare, cor

---

⁵⁹⁵ esse compositas, aut dependere ex his iam denotatis:] ex his iam recensitis componi, sive ab iis dependere. S1; sont dépendantes de celles-ci; C1.

⁵⁹⁶ humor] complexio S1; humeur C1.

⁵⁹⁷ humorem] temperamentum S1, 70; la [humeur] C1.

⁵⁹⁸ humor] Complexio S1; humeur C1.

⁵⁹⁹ timiditatem.] timor S1; timidité. C1.

⁶⁰⁰ Humor] Complexio S1; humeur C1.

⁶⁰¹ fortificatur] firmant S1; fortifient. Enfin, comme je viens de dire, la libéralité, la bonté, et l'amour dépendent de l'abondance des esprits, et forment en nous cette humeur qui nous rend complaisants et bienfaisants à tout le monde. La curiosité et les autres désirs dépendent de l'agitation de leurs parties; C1.

⁶⁰² § C1.

⁶⁰³ humores] complexiones S1; humeurs C1.

⁶⁰⁴ unde differentiae spirituum profluant.] differentiarum spirituum S1; d'où viennent les différences des esprits. C1, 60.

⁶⁰⁵ § 27 S1; § 57 C1.

⁶⁰⁶ fit commixtio,] recens permiscetur, S1; il se mêle tout fraîchement C1.

⁶⁰⁷ aliquo tempore post absolutam digestionem,] quidem postea: nimirum absoluta digestione, S1; quelque temps après que la digestion est achevée, C1.

⁶⁰⁸ § 58 C1.

⁶⁰⁹ aer respirationis] Aer insuper, qui inspiratur S1; L'air de la respiration C1.

⁶¹⁰ aliquando] certa ratione S1; en quelque façon C1.

⁶¹¹ involvat,] incendatur: S1; embrase C1.

⁶¹² § 59 C1.

perfecte disponit: unde fit ut spiritus⁶¹³ qui ex illo sanguine profluunt tunc⁶¹⁴,ᶜˣˡᵛⁱ sint in maiori copia et magis aequaliter agitati, et si contingat quod hepar illud⁶¹⁵ per nervos comprimetur, partes⁶¹⁶ subtiliores, quas continet ascendentes⁶¹⁷ ad cor etiam spiritus copiosores et magis vivos solito producent, [70] etiamsi non tam aequaliter agitatos. ⁶¹⁸Si autem fel, quod est destinatum ad purgandum sanguinem illarum partium quae ex omnibus magis sunt propriae ut involvantur⁶¹⁹,ᶜˣˡᵛⁱⁱ in corde, desinat fungi officio suo: aut si per nervum suum reservetur,⁶²⁰,ᶜˣˡᵛⁱⁱⁱ materia quam continet in venas redundat, spiritus erunt eo⁶²¹ magis vivi,⁶²² et simul magis inaequaliter agitati.

⁶²³Si splen, quaeᶜᶜˣˡᵛⁱ e contra destinatur ut purget sanguinem earum partium quae minus sunt propriae ut involventur⁶²⁴,ᶜˣˡⁱˣ in corde, sit male disposita,ᶜᶜˣˡᵛⁱⁱ aut si ita per suos nervos comprimetur, aut alio aliquo corpore⁶²⁵ ut materia quam continet in venas respiciat⁶²⁶,ᶜˡ spiritus erunt eo⁶²⁷ minus agitati,⁶²⁸ et etiam ⟦{in}⟧ minus inaequaliter agitati.⁶²⁹

⁶³⁰Denique omnia illa quae in sanguine aliquam mutationem possunt causare etiam aliquam in spiritus possunt: sed inter caetera,⁶³¹ nervus parvus qui terminatur in corde, cum possit dilatare et restringere tam ⟦amb⟧ ambos introitus, per quos sanguis vivorum⁶³²,ᶜˡⁱ [in margine: //]ᶜˡⁱⁱ et aer pulmonis descendit, quam

---

⁶¹³ unde fit ut spiritus] spiritus S1; les esprits C1.

⁶¹⁴ tunc] longe S1; d'autant C1.

⁶¹⁵ et si contingat quod hepar illud] Hepate vero S1; et s'il arrive que le foie C1.

⁶¹⁶ partes] partes sanguinis S1, 71; parties du sang C1.

⁶¹⁷ ascendentes] derepente (...) ascendentes S1; montant incontinent C1.

⁶¹⁸ § 60 C1.

⁶¹⁹ involvantur] inflammetur S1; embrasées C1.

⁶²⁰ reservetur,] restrictum, S1; resserré C1, 61.

⁶²¹ eo] tanto S1; d'autant C1.

⁶²² magis vivi,] vividiores, seu agiliores, S1; plus vifs, C1.

⁶²³ § 61 C1.

⁶²⁴ involventur] inflammari possunt S1; être embrasées C1.

⁶²⁵ alio aliquo corpore] aliove corpore S1; quelqu'autre corps que ce soit C1.

⁶²⁶ respiciat] regyrabit S1, 72; regorge C1.

⁶²⁷ eo] eo S1; d'autant C1.

⁶²⁸ minus agitati,] minus copiosi, minusque agitati, S1; moins abondants, et moins agités, C1.

⁶²⁹ minus inaequaliter agitati.] inaequalius acti S1; plus inégalement agités. C1.

⁶³⁰ § 62 C1.

⁶³¹ sed inter caetera,] sed omnium maxime S1; Mais par-dessus tout, C1.

⁶³² vivorum] venarum S1; des veines C1.

illos⁶³³ per quos sanguis ille exhalat se⁶³⁴ in arterias, potest causare mille differentias diversas in materia spirituum:⁶³⁵ sicuti calor aliquarum lampadum clausarum⁶³⁶ quibus alchimistae⁶³⁷ utuntur potest pluribus modis moderari [71] pro ut modo magis aut minus aperitur conclave,⁶³⁸ per quod oleum aut aliud alimentum flammae debet intrare, et modo illud, per quod sumus^cliii debet egredi. ⁶³⁹Quantum autem⁶⁴⁰,^cliv ad poros cerebri attinet non debent aliter a nobis imaginari, quam per modum aliquorum intervall[or]um quae intra⁶⁴¹,^clv fila alicuius texti⁶⁴² reperiuntur: totum enim cerebrum nihil aliud est, quam ⟦totum⟧ textum compositum aliquo modo particulari, quod hic vobis conabor explicare.

⁶⁴³Concipias igitur superficiem AA quae tanquam receptaculum aut textum satis densum et compressum⁶⁴⁴ concavitates E E respicit, unde^clvi omnes ....^clvii sunt⁶⁴⁵ tot [*in margine:* //]^clviii parvi tubi, per quos spiritus animales intrare possunt, et qui semper versus glandem FF⁶⁴⁶ respiciunt, unde spiritus illi, qui egrediuntur facile huc et illuc vertere se possunt versus diversa puncta istius glandis, uti alio modo vides esse versos in 48.ᵃ figura quam in 49.ᵃ.⁶⁴⁷,^clix Cogita etiam quod ex unaquaque parte istius receptaculi plurima fila valde gracilia egrediantur, unde⁶⁴⁸,^clx aliqua ordinarie sunt magis longa quam alia, et⁶⁴⁹ postquam illa fila se mutuo diversimode commiscuerunt⁶⁵⁰ per totum spacium notatum B,⁶⁵¹ longiora descen[72]dunt versus D et inde componentia medullam nervorum, se ipsa per⁶⁵² omnia membra spargunt. ⁶⁵³Etiam

---

⁶³³ illos] illos S1; les deux sorties C1.

⁶³⁴ exhalat se] exhalat S1; s'exhale et s'élance C1.

⁶³⁵ differentias diversas in materia spirituum:] differentias spiritibus S1; différences en la nature des esprits; C1.

⁶³⁶ lampadum clausarum] lampadum S1; lampes fermées C1.

⁶³⁷ alchimistae] chymicis S1; les alchimistes C1.

⁶³⁸ conclave,] illud orificium, S1; le conduit C1.

⁶³⁹ § 28 S1; § 63 and beginning of part 5, C1, 62.

⁶⁴⁰ Quantum autem] Quod vero S1; Secondement C1.

⁶⁴¹ intra] inter S1; entre C1.

⁶⁴² alicuius texti] lintei S1; de quelque tissu C1.

⁶⁴³ § 29 S1, 73; § C1, 63.

⁶⁴⁴ receptaculum aut textum satis densum et compressum] plexum satis expansum, densumque textum, cancellatim reticularum S1; plexum satis expansum, densumque textum, cancellatim reticulatum S2; une résille ou lacis assez épais et pressé C1.

⁶⁴⁵ unde omnes .... sunt] Cuius singula filamenta nihil aliud sunt S1; dont toutes les mailles sont C1.

⁶⁴⁶ FF] H S1; H C1.

⁶⁴⁷ in 48.ᵃ figura quam in 49.ᵃ] in figura 48 quam 49 S1; en la 48ᵉ qu'en la 49ᵉ figure, C1, 64; à l'endroit 48 qu'à l'endroit 49 C2, 57.

⁶⁴⁸ unde] Quorum S1; dont C1.

⁶⁴⁹ et] At vero S1; et C1.

⁶⁵⁰ commiscuerunt] reticulatim sunt intertexta S1; entrelacés C1.

⁶⁵¹ notatum B,] B, S1; marqué B, C1.

⁶⁵² se ipsa per] per S1; se C1.

⁶⁵³ § C1.

cogita qualitates praecipuas illorum filorum esse, satis facile in quamlibet partem[654] posse plicari[655] vi sola spirituum, qui ea tangunt, eodem fere modo ac si essent ex plumbo aut ex cera facta, retine[re] semper[656] ultima plica quae receperint,[657,ccxlviii] donec ipsis alia contraria[658] imprimatur. Cogita denique poros, de quibus hic,[659] nihil esse aliud quam intervalla quae reperiuntur in[660,clxi] his filis et possunt diversimode dilatari, et arctari vi spirituum qui intrant intus prout[661] est maior aut minor,[662] et breviora[663] reddere se in spatium LL[664] ubi unumquodque terminat se ad aliquam[665] ex naviculis[666,clxii] quae in eo sunt et ex iis recipit suam nutritionem.

[667]Ut[668,clxiii] autem commodius et facilius particularitates omnes[669] istius texti explicare possim, hic mihi incipiendum est loqui quomodo spiritus illi distribuantur.[670]

[671]Primo[clxiv] nunquam in aliquo loco quiescunt,[672] sed pro ut intrant concavitates cerebri E E per foramina parvae glandis notata FF[673] tendunt [73] statim versus canales A A qui illis magis directe opponuntur, et si canales isti A A non sint satis aperti ut recipiant res omnes,[674] ad minus saltem recipiant magis fortes, et magis vivos[clxv] suarum partium, dum debiliores et superflui[clxvi] reiiciuntur versus conductum[675,clxvii] IKL: qui respiciunt nares et pallatum, sc. magis agitatos[clxviii]

---

[654] in quamlibet partem] variis modis S1, 74; en toutes sortes de façons C1.
[655] plicari] plicari, flectique S1; être pliés C1.
[656] semper] adeoque semper S1; toujours C1.
[657] ultima plica] plicam (…) ultimo S1; les derniers plis C1.
[658] alia contraria] alia a spiritibus contrario modo agentibus S1; de contraires C1.
[659] hic,] hic questio, S1; ici question, C1.
[660] in] inter S1; entre C1.
[661] prout] quatenus nimirum illa vis, S1; selon qu'elle C1.
[662] minor,] remissior. S1; moins grande, et qu'ils sont plus ou moins abondants; C1.
[663] et breviora] breviora vero illius plexus filamenta S1; et que les plus courts de ces filets C1.
[664] LL] C C S1; c, c, C1.
[665] ad aliquam] ad extremitates quorundam S1; contre l'extrémité de quelqu'un C1.
[666] naviculis] vasculorum S1; petits vaisseaux C1.
[667] § 64 C1, 65.
[668] Ut] Ut S1; Troisièmement C1.
[669] commodius et facilius particularitates omnes] omnia S1; plus commodément (…) toutes les particularités C1.
[670] quomodo spiritus illi distribuantur.] de spirituum distributione. S1; de la distribution de ces esprits. C1.
[671] § C1.
[672] Primo nunquam in aliquo loco quiescunt,] Primo igitur, sese nulli loco alligant, S1; Jamais ils ne s'arrêtent un seul moment en une place; C1.
[673] FF] litera H S1, 75; H C1.
[674] res omnes,] omnes [spiritus] S1; tous [les esprits] C1.
[675] conductum] canales S1; les conduits C1, 66.

versus I, quam[676] quando multum ad huc[677,clxix] habent vigoris,[678] aliquando tanta vi[679,clxx] egrediuntur ut partes internas nasi titillent, quod causat sternutionem, deinde alia[e][680] versus K et L quo[681] facile possunt egredi, quia transitus sunt satis spaciosi, si vero errent[682] coacte redire[nt] versus A A canales, parvi qui sunt in superficie concava cerebri causant statim aliquam turbationem aut vertiginem, quae functiones imaginationis turbatur.

[683]Nota interim partes illas debiliores spirituum facile in pituitatem posse condensari,[684,clxxi] nunquam tamen sunt in cerebro nisi adsit aliquis morbus vehemens, sed in istis spaci⟦{…}⟧osis spaciis, quae sunt infra basin eius inter nares et guttur, eodem modo quo lumen[685] facile convertitur in fuliginem, in tubis cabinorum, nunquam tamen in foco in quo est ignis. [686]Nota etiam quod dum[687] spiritus egrediuntur ex glande FF[688] [74] tendant versus loca superficiei concavae[689] cerebri, quae iis magis directe sunt opposita, non intelligo[690] autem quod tendant versus illa quae iis in linea recta opposita sunt.[691,clxxii] [692]Cum[693] substantia cerebri sit valde mollis et plicabilis, concavitates[694] essent valde arctae, et prout apparet, in cerebro hominis mortui, fere omnes clausae nisi ⟦ten⟧ intrarent[695] aliqui spiritus: sed origo,

---

[676] I, quam] canalem I: per quem S1; I, par où C1.

[677] ad huc] adhuc S1; encore C1.

[678] vigoris,] vigoris (…), et exitum minus liberum inveniunt, S1; force, et qu'elles n'y trouvent pas le passage assez libre, C1.

[679] vi] impetuose S1; violence C1; violences C2, 59.

[680] deinde alia[e]] inde, reliqui spiritus se recipiunt S1; puis les autres C1.

[681] quo] in quos canales S1; par où C1.

[682] si vero errent] vel si illis ibidem introitus negetur S1; où si elles y manquent, C1.

[683] § C1.

[684] partes illas debiliores spirituum facile in pituitatem posse condensari,] particulas spirituum debiliores in pituitam facile condensari: S1; plus faibles parties des esprits, ne viennent pas tant des artères qui s'insèrent dans la glande H, comme de celles qui se divisant en mille branches fort déliées tapissent le fond des concavités du cerveau. Notez aussi qu'elles se peuvent aisément épaissir en pituite, C1.

[685] lumen] fumus S1; fumée C1.

[686] § C1, 67.

[687] dum] dum aio S1, 76; lorsque je dis que H C1.

[688] FF] H S1; H C1.

[689] concavae] concavae, sive interioris S1; intérieure C1.

[690] non intelligo] non hanc mihi mentem esse, ut intelligas S1; je n'entends pas C1.

[691] illa quae iis in linea recta opposita sunt.] illa duo, quae ipsis e diametro sunt opposita. S1; ceux qui sont vis-à-vis d'eux en ligne droite; mais seulement vers ceux, où la disposition qui est pour lors dans le cerveau les fait tendre. C1.

[692] § 65 C1.

[693] Cum] Certe quandoquidem S1; Or C1.

[694] concavitates] concavitates illius S1; ses concavités C1.

[695] nisi ⟦ten⟧ intrarent] nisi (…) se in illas reciperent S1; s'il n'entrait dedans C1.

quae spiritus illos producit est ordinarie ita copiosa, ut prout intrant illas concavitates,[696] ubi habent vires trudendi ab omni parte materiam[697] quae illos circuit, eamque inflare et[698] omnia parva fila nervorum, quae ex ea profluunt spandere, eodem modo quo ventus, aliquo modo vehemens, potest in⟦{ … }⟧flare vela alicuius navis[699] et omnes funes quibus affixae[clxxiii] sunt facere spandere. Unde fit ut pro tunc haec machina, cum sit disposita parere[700] omnibus actionibus spirituum, repraesentant[clxxiv] corpus hominis vigilantis: aut ad minus habent vires[701] trudendi, sic et spansam[clxxv] reddere aliquam partem,[702] dum aliae remanent liberae, ac inertes: sicuti faciunt vela alicuius navis quando ventus est nimis debilis ut illa inslet,[703,clxxvi] et tunc haec machina repraesentat corpus hominis dormientis, cui in somniis occurrunt diversas phan[75]tasmata. [704]Fingas igitur tibi ex. gr. quod ambae figurae M et N sit eadem quae est in cerebro hominis vigilantis, cum ea quae est in cerebro hominis dormientis,[705] et somniantis, [706]sed antequam de sumno[707] particularius agamus, debeo hic vobis dare consideranda omnia quae fiunt in cerebro tempore vigilationis, sc. quomodo formentur ideae obiectorum in loco destinato ad imaginationem, et quo ad sensum[708] communem quomodo reservent illum[709] in memoria,[clxxvii] et quomodo motum ⟦{earum}⟧ omnium membrorum[710] causent.

[711]Potes igitur videre in figura notata M quod spiritus qui ex glande FF[712] egrediuntur, cum partem cerebri notatam A dilataverint, et omnes poros aperuerint, inde currant versus B, deinde versus C, denique versus D, inde diffunduntur per omnes nervos, et hac ratione omnia fila, ex quibus nervi illi et cerebrum sunt composita,

---

[696] illas concavitates,] ventriculos cerebri S1; ces concavités, C1.

[697] materiam] obstantem sibi materiam S1; la matière C1.

[698] et] (…)que S1; et par ce moyen C1, 68.

[699] vela alicuius navis] vela S1; les voiles d'un navire, C1.

[700] parere] pariter obediat S1; à obéir C1.

[701] habent vires] spiritus vim habent S1; ils ont la force C1.

[702] partem,] cerebri partem S1; parties, C1.

[703] liberae, ac inertes: sicuti faciunt vela alicuius navis quando ventus est nimis debilis ut illa inslet,] immotis flaccidisque: uti quoque vela flaccida remissaque observamus, cum ventus debilior est, quam ut illa implere, planeque expandere possit. S1; libres et lâches, ainsi que font celles d'une voile, quand le vent est un peu trop faible pour la remplir C1, 69.

[704] § C1.

[705] quod ambae figurae M et N sit eadem quae est in cerebro hominis vigilantis, cum ea quae est in cerebro hominis dormientis,] eam esse differentiam, inter M et N, quam inter vigilantem et dormientem, S1, 77; que la différence qui est entre les deux figures M, et N, est la même qui est entre le cerveau d'un homme qui veille, et celui d'un homme qui dort, C1.

[706] § C1.

[707] sumno] somno eiusque insomniis S; sommeil et des songes C1, 69–70.

[708] et quo ad sensum] atque sensui S1; et pour le sens C1.

[709] [ideae] reservent illum [sensum communem]] illae [ideae] (…) conserventur S1, 78; elles [idées] se réservent C1, 70.

[710] ⟦{earum}⟧ omnium membrorum] omnium membrorum S1; de tous les membres C1.

[711] § 30 S1; § 66 C1.

[712] FF] H S1; H C1.

tenent ita tensa, ut actiones quae habent quantumvis parvas vires, ea movendi, communicent se facile ab una earum extremitatum usque ad alteram absque eo quod impedimenta viae, quam transeunt, eos impediant. [713]Ut autem illa impedimenta non impediant te, etiam quo minus clare videas, quomodo hoc etiam [76] inserviat,[714] ad formandas ideas obiectorum, quae afficiunt sensus. Vide in figura sequenti[715] parva fila 12, 34, 56,[716] et similia, quae nervos opticos componunt, et sunt sparsi a fundo oculi 1 3 5,[717] et cogita quod fila illa ita sint disposita, ut si radii, qui v. gr. veniunt a puncto A[718] comprimant fundum oculi in puncto I[719] trahunt ea ratione omnia fila 12[720] et augent aperturam canalis parvi[clxxviii] notati[721] 2 eodem prorsus modo quo radii venientes a puncto S[722] augent aperturam canalis parvi notati[723] 4 et sic de caeteris. Ita ut diversi modi quibus puncta 1 3 5 istis radiis comprimuntur del[in]eant in fundo oculi figuram quae referat se figurae obiecti[724] ABC sicut supra dictum est, evidens enim est modos diversos quibus canales parvi 2 4 6 sunt aperti filis 12, 34, 56,[725] etc. eam[726] debere etiam del[in]e[a]re in concava superficie cerebri. [727]Deinde cogita quod spiritus qui conantur intrare in unoquoque ex illis tubis parvis,[728] et similes, non veniant indifferenter ex omnibus illis punctis, quae sunt in superficie glandis FF,[729] sed tantum ex aliquibus in particulari, et quod ii qui ex. gr. veniunt ex puncto A sint qui conantur intrare in tubis[730,clxxix] 4 et 6 et sic de caeteris.

[77] [clxxx]Ita ut instant[i] primo, quo apertura illorum canalium sit maior, spiritus incipiunt egredi liberius et citius, quam antea faciebant sc. per puncta[731] superficiei istius glandis quae eos respicit, et sicuti diversi modi quibus canales 2 4 6 sunt ap[er]ti,

---

[713] § 67 Cl.

[714] hoc etiam inserviat,] illa obliquitas inserviat S1, 79; cela sert C1.

[715] Vide in figura sequenti] oculos in hanc figuram coniice. Vide in sequenti figura S1; regardez en la figure ci-jointe C1, 71.

[716] 12, 34, 56,] 1, 2, 3, 4, 5, 6: S1; 1 2, 3 4, 5 6, C1.

[717] 1 3 5,] 1, 3, 5, usque ad interiorem cerebri superficiem 2, 4, 6 et S1; 1, 3, 5, jusques à la superficie intérieure du cerveau 2, 4, 6. C1.

[718] A] A S1, 80; A de l'objet C1.

[719] I] i S1; 1 C1.

[720] omnia fila 12] totum filamentum 1, 2 S1; tout le filet 1 2 C1.

[721] aperturam canalis parvi notati] tubulum notatum numero S1; l'ouverture du petit tuyau marqué C1.

[722] venientes a puncto S] incidentes in punctum 3 S1; qui viennent du point B C1.

[723] aperturam canalis parvi notati] tubulum S1; l'ouverture du petit tuyau C1.

[724] figurae obiecti] obiecto S1; à celle de l'objet C1.

[725] 12, 34, 56,] 1, 2, 3, 4, 5, 6, S1; 1 2, 3 4, 5 6, C1.

[726] eam] sui quoque vestigium S1; la C1.

[727] § 31 S1; § 68 C1, 72.

[728] tubis parvis,] tubulos 2, 4, 6, S1; petits tuyaux 2, 4, 6, C1.

[729] FF,] H, S1; H, C1.

[730] A sint qui conantur intrare in tubis] A, conniti, ut ingrediantur tubos S1; a, de cette superficie, qui tendent à entrer dans le tuyau 4 [*errata corrige:* tuyau 2], et ceux des points b, et c, qui tendent à entrer dans les tuyaux C1.

[731] puncta] puncta S1; endroits C1.

del[in]e[a]nt figuram unam, quae referat se figurae obiecti[732] ABC in superficie concava cerebri sicuti ille[clxxxi] quibus spiritus[733] egrediuntur ex punctis A B C eam del[in]e[a]nt in superficie istius glandis.

[734]Nota autem me per figuras[735] hic non intelligere tantum illa,[736] quae aliquando repraesentant positionem linearum superficierum, et obiectorum,[737] sed et[iam] omnia illa quae secundum ea quae supra dixi possent animae dare[738] percipere motum, distantiam, magnitudinem, colores, sonos, odores, et alias similes qualitates, imo illas quae darent animae percipere titillationem, dolorem, famem,[739] laetitiam, tristitiam, et alias similes passiones. Facile enim est intellectu[740] quod tubus 2 verbi gr. alio modo aperietur per actionem, quam dixi causare sensationem titillationis,[741] quam per illam quam dixi causare sensationem coloris albi, aut etiam illam [78] doloris, et quod spiritus qui egrediuntur a puncto A[742] diversimode versus hunc canalem tendent, prout diversimode erit ap[er]tus.[743] [744]Inter illas figuras autem non sunt illae quae imprimuntur in organis sensuum exteriorum aut in superficie interna cerebri, sed tantum illae quae in spiritibus del[in]e[a]ntur[745] in superficie glandis FF[746] ubi sedes imaginationis et sensus communis est posita, quae debent capi pro ideis hoc est pro formis, aut imagin[[...}]]ibus, quos[clxxxii] anima rationalis immediate considerabit, ubi cum isti machinae unietur illa imaginabit sibi aut percipiet aliquod obiectum. [747]Nota autem me hic dicere imaginabitur, aut percipiet,[748,clxxxiii] quia sub notione ideae generaliter comprehendere volo impressiones omnes quae egrediendo ex glande FF[749] recipere possunt spiritus, quae omnes eam tribuunt sensibus communibus,[750] quando debent ex praesentia obiectorum,

---

[732] figurae obiecti] obiecto S1; à celle de l'objet C1.

[733] sicuti ille quibus spiritus] Sic etiam illi modi determinationesve, quae oriuntur a spiritibus S1; ainsi celle [figure] dont les esprits C1.

[734] § 69 C1.

[735] figuras] loco figurae nomine S1, 81; ces figures C1.

[736] illa,] illa S1; les choses C1.

[737] positionem linearum superficierum, et obiectorum,] linearum determinationes, obiectorumque superficies S1; la position des lignes et des superficies des objets, C1.

[738] animae dare] animae occasionem exhibere possunt S1; pourront donner occasion à l'âme C1.

[739] famem,] famem, S1; la faim, la soif, C1.

[740] facile enim est intellectu] Minime enim obscurum est, S1; Car il est aisé à entendre, C1.

[741] sensationem titillationis,] sensum coloris rubri, vel sensum titillationis: S1; le sentiment de la couleur rouge, ou celui du chatouillement, C1, 73.

[742] A] A S1; a C1.

[743] diversimode ap[er]tus.] diversimode fuerit reseratus: et sic de caeteris. S1; sera ouvert diversement, et ainsi des autres C1.

[744] § 70 C1.

[745] in spiritibus del[in]e[a]ntur] afficiuntur spiritus S1; se tracent dans les esprits C1.

[746] FF] H S1; H C1.

[747] § 71 C1.

[748] imaginabitur, aut percipiet,] imaginabitur, aut sentiet: S1; imaginera, ou sentira; C1.

[749] FF] H S1, 82; H C1.

[750] sensibus communibus,] ad sensum communem S1; au sens commun, C1.

cum tamen etiam ex aliis⁷⁵¹ causis procedere possint, de quibus postea dicam, et tunc illae debent imagin⟦{...}⟧ationi tribui. ⁷⁵²Possem hic etiam addere quomodo lineae⁷⁵³ istarum idearum transeant per arterias versus cor, et sic per totum sanguinem radient,⁷⁵⁴ imo quomodo possint aliquando certis [79] actionibus matris determinari imprimere se in membris embrionis,⁷⁵⁵,ᶜˡˣˣˣⁱᵛ qui formatur in [*in margine: //*]ᶜˡˣˣˣᵛ visceribus matris,⁷⁵⁶ sed sufficit mihi vobis dicere adhuc quomodo imprimant se in partibus internis cerebri notatis B ubi est sedes memoriae. ⁷⁵⁷Cogita igitur⁷⁵⁸ quod postquam spiritus qui egrediuntur ex glande FF⁷⁵⁹ receperint impressionem alicuius ideae transeant inde per canales 2 4 6 et similes in poros aut intervalla quae sunt in⁷⁶⁰,ᶜˡˣˣˣᵛⁱ filis parvis ex quibus haec pars cerebri⁷⁶¹ est composita, et habeant vires haec intervalla aliquantisper dilatare, plicare, et diversimode disponere fila parva quibus occurrunt⁷⁶²,ᶜˡˣˣˣᵛⁱⁱ secundum modos diversos quibus moventur et diversos canales⁷⁶³ per quos transeunt. Ita ut ibi etiam del[in]eant figuras quae referant se⁷⁶⁴ figuris obiectorum: non tamen adeo facile, nec adeo perfecte⁷⁶⁵ quam in glande FF⁷⁶⁶ sed paulatim melius et melius⁷⁶⁷ prout eorum actio est fortior aut diutius durat, aut prout saepius est retracto.⁷⁶⁸,ᶜˡˣˣˣᵛⁱⁱⁱ Qua de causa figurae illae non delenterᶜˡˣˣˣⁱˣ tam facile sed ita conservantur, ut ipsis mediantibus ideae quae aliquando fuerint in ista glande, diu post iterum possint formari absque [80] eo quod praesentia obiectorum, quibus referuntur ad hoc, sit requisita.⁷⁶⁹ ⁷⁷⁰Quando v. gr. actio obiecti ABC quae

---

⁷⁵¹ aliis] multis aliis S1, 83; plusieurs autres C1.

⁷⁵² § C1.

⁷⁵³ lineae] vestigia S1; les traces C1.

⁷⁵⁴ radient,] radios suos diffundant. S1; rayonnent C1.

⁷⁵⁵ embrionis,] infantis, S1; enfant C1.

⁷⁵⁶ in visceribus matris,] in matris utero S1; dans ses entrailles. C1.

⁷⁵⁷ § 72 C1, 74.

⁷⁵⁸ igitur] igitur S1; donc à cet effet C1.

⁷⁵⁹ FF] H S1; H C1.

⁷⁶⁰ in] inter S1; entre C1.

⁷⁶¹ pars cerebri] cerebri pars B B S1; partie du cerveau B C1.

⁷⁶² occurrunt] transeuntes impingunt, S1; rencontrent en leurs chemins C1; rencontrent en leur chemin C2, 67–68.

⁷⁶³ canales] aperturas tubulorum S1; ouvertures des tuyaux C1.

⁷⁶⁴ quae referant se] similes S1; qui se rapportent C1.

⁷⁶⁵ adeo facile, nec adeo perfecte] primo, ictu tanta facilitate aut perfectione S1; si aisément ni si parfaitement du premier coup C1.

⁷⁶⁶ FF] H S1; H C1.

⁷⁶⁷ melius et melius] melius, accuratiusque S1, 84; mieux en mieux, C1.

⁷⁶⁸ retracto.] repetita. S1; réitérée. C1, 75.

⁷⁶⁹ absque eo quod praesentia obiectorum, quibus referuntur ad hoc, sit requisita.] absentibus etiam obiectis, quorum simulacra sunt (...). Qua in re consistit memoria. S1; sans que la présence des objets auxquels elles se rapportent y soit requise. Et c'est en quoi consiste la mémoire. C1.

⁷⁷⁰ § C1.

auget aditum[771] ad canales 2 4 6 8,[772] est in causa cur intus intrent in maiori copia quam alias facerent absque hoc, est etiam in causa cur dum transeunt per N[773] habeant vires certos aditus sibi formare, qui remanent[774] adhuc postquam actio obiecti ABC cessavit, aut si iterum claudantur certam aliquam dispositionem in parvis[775] relinquunt ex qua[776,cxc] haec pars cerebri N est composita, mediante qua possunt multo facilius iterum aperiri, quam si nondum fuissent aperti. Eodem modo s⟦{...}⟧i plures acus[777] aliquis transiret per aliquam telam,[778] uti vides in ea quae est notata A parva foramina quae in ea fierent remanerent adhuc ap[er]ta, uti versus[779] postquam acus ille[cxci] essent retractae, vel si iterum clauderentur relinquerent tamen in illa tela ⟦{q}⟧ aliqua signa[780] sicuti versus O[781] quae essent causae cur[782] facile possent de novo aperiri, [783]uno[784,cxcii] notandum est quod si tantum⟦{a}⟧ aliqua ex illis aperirentur uti A et B illa sola ratione possent alia,[785] sicuti C et D etiam eodem tempore aperiri [81] praecipue si simul saepius fuissent aperta, et aliqua sine aliis non solita fuissent ⟦{...}⟧ aperiri, quod etiam ostendit[786] quod alicuius rei recordatio possit per aliam rem excitari, quae aliquando eodem tempore in memoria simul sint impressa, et si duos oculos[787] cum naso videam statim imaginor me etiam videre frontem et os, et simul alias omnes partes faciei, ex eo quod illa solebant semper esse coniuncta simul, nec soleam unum videre sine altero.[788] Et videns ignem statim recordor sui caloris, quia illam[cxciii] sensi saepius dum viderem ignem. [789]Considera ulterius glandem A[790] esse compositam materia valde molli, quae non est omnino unita substantiae cerebri,

---

[771] auget aditum] magis aperit S1; augmentant l'ouverture C1.

[772] 2 4 6 8,] 2, 4, 6, 8, S1; 2, 4, 6, C1.

[773] transeunt per N] ulterius transeunt ad N S1; passant plus outre vers N C1.

[774] remanent] remaneant aperti S1, 85; demeurent ouverts C1.

[775] parvis] filamentis S1; petits filets C1.

[776] ex qua] ex quibus S1; dont C1.

[777] acus] acus S1; aiguilles, ou poinçons C1.

[778] aliquis transiret per aliquam telam,] transiissent textum, S1; on passait (...) au travers d'une toile. C1.

[779] versus] circa B S1; vers a et vers b C1.

[780] tamen (...) ⟦{q}⟧ aliqua signa] quaedam nihilominus signa S1; des traces C1.

[781] versus O] circa O S1; vers c et vers d C1.

[782] quae essent causae cur] quae post S1; par le moyen de laquelle C1.

[783] § 73 C1, 76.

[784] uno] imo S1, 86; Et même C1.

[785] alia,] etiam alia, S1; les autres, C1.

[786] quod etiam ostendit] Unde liquet S1; Ce qui montre C1.

[787] duos oculos] oculisque S1; deux yeux C1.

[788] illa solebant semper esse coniuncta simul, nec soleam unum videre sine altero.] nullam illarum partium videre solitus sum, quin simul omnes intuear. S1; je n'ai pas accoutumé de les voir l'une sans l'autre. C1.

[789] § 74 C1, 77.

[790] A] H S1, 87; H C1.

sed solum arteriis parvis affixa, quorum pelles sunt satis amplae[791] et plicabiles, et quasi per modum bilancis substantiae[792,cxciv] vi sanguinis, quem calor ignis trudit versus ipsam, ita ut non multa requirantur ad determinandam ipsam inclinare se et protendere plus aut minus modo in unam partem, modo in aliam, et facere ut se inclinando disponat spiritus qui ex e⟦{...}⟧a egrediuntur ut currant versus aliquos[cxcv] partes cerebri potius quam versus alias.

[82] [793]Sunt autem hic duo[794] principalia, absque eo quod numeramus vires animae (de quibus postea dicam) quae eam[795] sic possunt movere, et quae mihi hic sunt explicanda, [796]primum igitur est differentia quae occurrit inter partes parvas spirituum, qui egrediuntur ex ipsa: si enim ⟦omnes il⟧ omnes illi spiritus exacte haberent vires aequales, nec esset ulla alia causa quae determinaret ipsam inclinare se, nec illuc[797] fluerent aequaliter per omnes eius[798] poros, eamque sustine⟦...⟧nt rectam omnino et immobilem in centro capitis, sicut ostensa est in figura 40.ª, sed tanquam corpus aliquod, affixum solum aliquibus filis, quod sustineretur vi fumi qui egreditur ex fornace, fluctuaret semper[799] pro ut partes diversae istius fumi diversimode in illud agitarent, sic etiam partes parvae istorum spirituum, quae glandem istam sublevant atque sustinent, cum sint fere semper indifferentes ad aliquid, eam agitant et inclinant,[800] modo in unam partem, modo in aliam pro ut videre potes in figura 41,[801,cxcvi] ubi non tantum eius centrum FF[802] porum[cxcvii] distat a centro cerebri notato O,[803] sed etiam extremitates arteriarum quae eam sustinent, [83] sunt ita incurvatae ut fere omnes spiritus, quos ipsi adferunt capiant cursum per partem superficiei ABC versus canales parvos[cxcviii] 2 4 6 8[804] aperientes hac ratione poros suos[805] qui versus partem illam[806] magis[807] respiciunt, quam versus aliam.[808] [809]Praecipuus autem effectus qui egreditur ex his, in eo consistit, quod spiritus egredientes sic magis particulariter

---

[791] amplae] laxae S1; lâches C1.
[792] substantiae] suspensa S1; soutenue C1.
[793] § 75 C1.
[794] duo] duo S1, 88; deux causes C1.
[795] eam] glandulam illam S1; la C1.
[796] § C1.
[797] nec illuc] huc, illucve S1; ni çà ni là C1, 78.
[798] eius] glandulae S1; ses C1.
[799] semper] continuo S1; incessamment çà et là C1.
[800] agitant et inclinant,] inclinant. S1; ne manquent pas de l'agiter et faire pencher C1.
[801] 41,] 41. S1; 41, ["4" in the figure itself] C1; 4, C2, 71.
[802] FF] H S1; H C1.
[803] O,] litera O: S1; O, C1.
[804] 2 4 6 8] 2, 4, 6, 8 S1, 89; 2, 4, 6 C1.
[805] poros suos] illos illorum tubulorum poros S1; ceux de ses pores C1.
[806] versus partem illam] eo S1; vers là C1.
[807] magis] directius S1; beaucoup davantage C1.
[808] quam versus aliam.] caeteris S1; que les autres. C1.
[809] § 76 C1.

ex aliquibus partibus[810] quam ex aliis possint habere vires vertendi canales parvos superficiei concavae cerebri, in quibus se recipiunt, versus partes[811] ex quibus egrediuntur nisi iam omnes esse conversos reperiant, hac ratione vertendi membra quibus referunt se canales isti versus loca quibus referunt se partes illae superficiei glandis FF.[812] Nota etiam ide[a]m motus membrorum non consistere in modo alio quam illo quo spiritus pro tunc egrediuntur ex illa glande, et quod sic eius idea illos[cxcix] causet,[813] [814]pro ut verbi gr. supponi potest hic, quod hoc quod facit ⟦ad⟧ ut tubus[815] 8 vertat se potius versus punctum {6}[816] quam versus aliquod [84] aliud, hoc fit ex eo quod[817] spiritus qui ex illo puncto egrediuntur maiori vi tendant in illud quam aliud, et quod hoc ipsum esset in causa cur anima[818] perciperet brachium vertere se versus B[819] si hoc ipso esset in illa machina, prout supponam postea illam esse,[820,cc] cogitandum est enim quod omnia puncta[821] versus quae tubus ille[822] versus esse potest, ita correspondeant omnibus locis versus quae brachium notatum 7 esse[823] potest ut quod iam facit quod brachium sit versum versus obiectum B est ex eo, quod tubus ille respiciat punctum {6}[824] quod si iam spiritus mutantes cursum verterent tubum hunc versus aliquod aliud punctum[825] uti versus C fila parva 7[826] quae cum egrediantur ex eius vicinia in membris istius ⟦{...}⟧brachii[827] se recipiunt, mutantia illa, eadem ratione situm restringerent aliqua[cci] ex poris cerebri[828] quae sunt versus D[ccii] et dilatarent aliqua alia,[cciii] qua ratione spiritus inde in istos muscu[85]los transeuntes alio modo quam nunc faciant statim brachium versus obiectum C verterent,[829,cciv]

---

[810] partibus] partibus superficiei glandis S1; endroits de la superficie de cette glande C1.

[811] partes] loca S1; endroits C1.

[812] glandis FF.] H. S1, 90; de la glande H. C1, 79.

[813] eius idea illos [spiritus] causet,] motum ipsum a sua idea formari. S1; c'est son idée qui le cause. C1.

[814] § 77 C1.

[815] quod hoc quod facit ⟦ad⟧ ut tubus] causam, qua tubus S1; que ce qui fait que le tuyau C1.

[816] {6}] b S1; b C1.

[817] hoc fit ex eo quod] causam (...) illam duntaxat esse, quod S1; c'est seulement que C1, 80.

[818] esset in causa cur anima] animae occasionem exhibet S1; donnerait occasion à l'âme C1.

[819] B] B S1; l'objet B C1.

[820] si hoc ipso esset in illa machina, prout supponam postea illam esse,] si sic machinae brachium sit dispositum, uti modo supponam. S1; si anima huic machinae insit, uti modo supponam. S2; si elle était déjà dans cette machine, ainsi que je l'y supposerai ci-après. C1.

[821] puncta] puncta S1; les points de la glande C1.

[822] tubus ille] tubi illi S1; ce tuyau 8 C1.

[823] esse] moveri S1; être C1.

[824] {6}] b S1; b de la glande C1.

[825] punctum] punctum S1; point de la glande C1.

[826] 7] 7 S1; 8, 7 C1.

[827] in membris istius ⟦{...}⟧brachii] in ea brachii membra S1; dans les muscles de ce bras C1.

[828] restringerent aliqua ex poris cerebri] quosdam cerebri poros S1; quosdam cerebri poros (...) angustiores redderent S2; rétréciraient quelques-uns des pores du cerveau C1.

[829] verterent,] convertant S1, 92; converterent S2; tourneraient C1.

pro ut reciproce si aliqua alia actio quam actio spirituum qua[830] intra[n]t per tubos 8[831,ccv] illud idem brachium verteret versus B vel versus C faceret ut isti tubi 8 verterent se versus 6 aut versus [[{...}]]C,[832] ita ut idea istius modus[ccvi] se etiam eodem tempore formaret, modo attentio non sit turbata, hoc est si glans FF[833] non sit impedita quo minus se inclinet versus 8 aliqua alia actione fortiori, sic etiam generaliter cogitandum est quod quilibet alii tubi parvi qui sunt in superficie concava cerebri referant se quibuslibet aliis membris, et quaelibet alia puncta superficiei glandis FF[834] quibuslibet partibus,[835] versus quas membra illa possunt esse conversa, ita ut motus istorum membrorum et ideae ipsorum possint se mutuo reciproce causare. [836]Et quod magis est occasione istorum [*subscriptum:* ∧][ccvii] intelligen[86]dumque[837,ccviii] nobis est, non modo[838] dum ambo oculi istius machinae, et organa plurium aliorum eius sensuum sunt directi versus aliquod idem obiectum, quod non ideo formentur plures ideae in suo cerebro, sed[[{...}]] tantum unica: cogitandum [*subscriptum:* ∧; *superscriptum:* enim][ccix] est[839] semper ex illis iisdem punctis istius superficiei glandis FF[840] egredi spiritus, qui cum in diversos tubos tendant, possunt diversa membra versus diversa[841] obiecta vertere, pro ut[842] ex solo puncto 6[843] egrediuntur spiritus qui cum tendant in tubos 4, 4, et 8, eodem tempore vertunt ambos oculos et brachium dextrum in obiectum B, [844]quod facile erit credendum, ut etiam intelligatur in quo consistat idea distantiae obiectorum, [[con]] cogitandum est quod prout superficies mutet situm[845] iidem suorum punctorum referunt se aliis locis eo magis distantibus a cerebro[846] notato O quo[847] sunt magis propinqui, [[quo]] et eo magis propinqui quo ab illo sunt [87] magis remoti, prout hic ex. gr.[848] si si [*sic*] punctum

---

[830] qua] qui S1; qui C1.

[831] tubos 8] tubulum 8 S1; tubulos 87 S2; le tuyau 8 C1.

[832] isti tubi 8 verterent se versus 6 aut versus [[{...}]]C,] isti tubi 8 aut d, sese verterent ad b, aut c: S1; ce tuyau 8 se tournerait vers les points de la glande b ou c. C1.

[833] FF] H S1; H C1.

[834] FF] H S1; H C1, 81.

[835] partibus,] locum S1; côtés C1.

[836] § 78 C1.

[837] occasione istorum [*subscriptum:* ∧] intelligendumque] hac occasione intelligimus S1; entendre ici par occasion C1.

[838] non modo] quare S1; comment C1.

[839] cogitandum [*subscriptum:* ∧; *superscriptum:* enim] est] Cogitandum est S1; il faut de penser C1, 82.

[840] FF] H S1; H C1.

[841] diversa] diversa S1; les mêmes C1.

[842] pro ut] Quemadmodum hic S1, 93; comme ici que c'est C1.

[843] 6] b S1; b C1.

[844] § 79 C1.

[845] superficies mutet situm] superficies situm mutat S1; superficies haec situm mutat S2; cette superficie change de situation C1.

[846] cerebro] centro cerebri S1; centre du cerveau C1.

[847] quo] quanto S1; que ces points C1.

[848] hic ex. gr.] hic, exempli gratia S1; ici il faut penser que C1.

6^{849} paulisper retrocessisset, quod non fit,^{850,ccx} referret se loco magis distanti quam fit^{ccxi} B et si aliquantisper esset inclinatum versus anteriorem partem referret se uno magis vicin⟦i⟧o. ^{851}Et hanc ratione dum erit anima ⟦ha⟧ [*superscriptum:* in]^{ccxii} hac machina,^{852,ccxiii} illa aliquando poterit diversa obiecta percipere mediantibus iisdem organis e⟦a⟧odem modo dispositis, et absque eo quod quidquam mutetur praeter situm glandis FF^{853} pro ut hic ex. gr. anima poterit percipere quod est in puncto L mediantibus ambabus manibus quae tenent utrosque baculos NL et OL ex eo quod ex puncto L glandis FF^{854} sit quod egrediantur spiritus qui vadunt versus 7 et versus 8,^{855} quibus correspondent illae^{856,ccxiv} duae manus. Ita ut puncta superficiei^{857} N et O essent in locis notatis I et K et per consequens quod ex illis egrederentur spiritus qui vadunt^{858} versus 7 et 8 [88] anima percipere deberet illud quod est versus N et O mediantibus iisdem manibus, et absque eo quod in alio sint.^{859} ^{860}Notandum tamen est quod^{861} dum^{ccxv} glans FF^{862} est inclinata versus partem aliquam sola vi spirituum et absque eo quod anima rationalis, nec sensus externi aliquid contribuant, ideae quae formantur in superficie,^{863} non procedant solum ex inaequalitatibus quae occurrunt in partibus parvis^{864,ccxvi} istorum spirituum, et quae causant differentiam humorum,^{865} pro ut supra fuit dictum, verum etiam ex impressionibus memoriae. Si enim figura alicuius obiecti particularis sit impressa multo distinctius quam quaelibet alia in parte cerebri versus quam praecise est inclinata haec glans, spiritus qui etiam eo tendunt, non possunt aliter quin etiam recipiant impressionem, et hac ratione res praeteritae redeunt aliquando mere fortuito, in cogi[89]tationem, absque eo quod memoria

---

[849] 6] B S1; b C1.

[850] retrocessisset, quod non fit] retro cessisset S1; était un peu plus retiré en arrière qu'il n'est pas C1.

[851] § 80 C1.

[852] ⟦ha⟧ [*superscriptum:* in] hac machina,] machinae S1; dans cette machine, C1.

[853] FF] H S1; H C1.

[854] FF] H S1; H C1.

[855] vadunt versus 7 et versus 8,] tendunt ad 7 et ad 8, S1; entrent dans les tuyaux 7, et 8, C1.

[856] illae] illae S1; ses C1, 83.

[857] Ita ut puncta superficiei] Si vero haec glans H esset paulo vicinior, quam etiamnum est ita; ut puncta 93–94; au lieu que si cette glande H, était un peu plus en avant qu'elle n'est, en sorte que les points C1.

[858] vadunt] tendunt S1, 94; vont C1.

[859] absque eo quod in alio sint.] neutiquam quoque mutatis. S1, 95; et sans qu'elles fussent en rien changées. C1.

[860] § 81 C1, 84.

[861] Notandum tamen est quod] Notandum tamen est S1; Denique notandum S2; Au reste, il faut remarquer que C1.

[862] FF] H S1; H C1.

[863] superficie,] glandis superficie S1; sa superficie C1.

[864] partibus parvis] particularium S1; minimarum particularium S2; les petites parties C1.

[865] humorum,] complexionis S1; humeurs C1.

sit valde excitata, aliquando[866],[ccxvii] obiecto quod tangat sensus [867]si vero plures figurae et diversae reperiantur del[in]e[a]tae in eadem partem cerebri, faere[ccxviii] aeque perfecte una quam altera, pro ut saepe contingit, spiritus aliquid recipient de impressione uniuscuiusque et hos[ccxix] plus aut minus secundum diversas occurrentias earum partim. Et eodem modo componuntur chymerae, et hippocentauri[868] in imaginatione illorum qui vigilando dilirant, hoc est qui eorum phantasias pusillanime permittunt disvagare, absque eo quod obiecta externa ipsam divertant, nec eorum iudicio patiantur repulsam.[869] [870]Verum effectus memoriae qui mihi magis dignus hic considerari videtur in eo consistit, quod absque eo quod sit aliqua ea[871],[ccxx] in ista machina possunt naturaliter disponi, ut imitet motus quos homines veri, aut aliae similes machinae[872] facerent ibi praesentes.

[873]Causa secunda quae motus glandis FF[874] determinare potest, est actio obiectorum [90] quae tangunt sensus, facile enim est intellectu, quod introitus parvorum tuborum[ccxxi] 2 4 6 8[875] verbi gr. cum sit dilatatus per actionem obiecti ABC, spiritus qui statim fluere incipiunt, liberius et celerius quam faciebat, attrahunt secundum aliquid ex ista glande, et faciunt ut se inclinet, si aliunde non impediatur, et mutans dispositionem suorum pororum incipit conducere maiorem numerum spiritum per A B C versus 2 4 6 8[876] quam antea faciebat, qua ratione idea quae formatur ab illis spiritibus fit eo perfectior, et in eo consistit effectus primus qui notandus est.

[877]Secundus consistit in eo quod tempore quo glans illa detinetur sic incurvata in aliquam partem, hoc eam impedit quo minus tam facile possit recipere ideas obiectorum, quae agunt in organa aliorum sensuum, pro ut hic ex. gr. tempore quo omnes fere spiritus, quos producit glans FF,[878] [91] egrediuntur ex punctis A B C non egrediuntur sufficienter a puncto E[879] ut formare possint ideam obiecti D unde[880],[ccxxii] actionem non esse adeo vivam[881],[ccxxiii] nec adeo fortem, quam actionem ABC

---

[866] aliquando] a nullo S1; par aucun C1.

[867] § 82 C1.

[868] hippocentauri] hippogryphi S1; hippogriffes C1.

[869] nec eorum iudicio patiantur repulsam.] nullave ratione gubernant, sed laxatis habenis S1, 96; ni qu'elle soit conduite par leur raison. C1.

[870] § 83 C1.

[871] ea] anima S1; ame C1, 85.

[872] similes machinae] quoque rerum similium S1; semblables machines, C1.

[873] § 84 C1.

[874] FF] H S1; H C1.

[875] 2 4 6 8] 2, 4, 6, 8 S1, 97; 2, 4, 6, C1.

[876] 2 4 6 8] 2, 4, 6, 8, S1; 2, 4, 6, C1, 86.

[877] § 85 C1.

[878] FF,] H S1; H, C1.

[879] E] d S1; d C1.

[880] unde] Unde S1; Cuius S2; dont C1.

[881] vivam] vivam S1; vividam S2; vive C1.

suppono:[882,ccxxiv] ex quibus vides ideas se mutuo impedire[883,ccxxv] et inde fit quod eodem tempore non possimus esse attenti pluribus rebus.[884] [885]Etiam notandum est quod dum organa sensuum incipiunt tangi per aliquod fortius quam per alia,[886,ccxxvi] cum non sint tam integre disposita quam esse possent,[887] praesentia istius obiecti sufficiens est, ut absolvat ea disponat.[888] Et sic oculus ex. gr. sit dispositus locum aliquem remotum respicere dum obiectum ABC quod est valde propinquum incipit fieri praesens ipsi, dico quod actio istius obiecti poterit illum determinare, ut illud oculis fixis statim respiciat. [889]Ut autem hoc sit facilius intellectu, primo considera differentiam, quae est inter oculum dispositum respicere obiectum remotum, uti est in 50.^{a} figura, quae parum est [92] magis accurata, et inter alias partes oculi[890] proportionaliter alio modo dispositas, in hac ultima[891] figura quam in praecedenti:[ccxxvii] verum etiam quia parvi tubi 2 4 6 magis inclinati sunt imo uno puncto propinquiori, et quia glans FF[892] parum magis est ipsos[ccxlix] approximata, et [[quia]] quia pars, sive locus[893] superficiei ABC est proportionaliter parum magis curvatus,[894] ita ut tam in una figura quam in altera semper a puncto A egrediuntur spiritus, qui tendunt in tubum 4 et a puncto C quod egrediuntur illi qui tendunt in tubumB.[895] [896]Etiam considera solos illos motus glandis FF[897] esse sufficientes ut mutent situm istorum tuborum, et consequenter dispositionem corporis oculi,[898] pro ut generalibus terminis supra fuit dictum, illos posse cogere ad motum omnium membrorum.

---

[882] suppono:] concludo S1; suppono S2; je suppose C1.

[883] ideas se mutuo impedire] ideas sibi mutuo obesse S1; quomodo ideae sibi mutuo obsint S2; comment les idées s'empêchent l'une l'autre C1.

[884] eodem tempore non possimus esse attenti pluribus rebus.] pluribus intentis, minor est ad singula sensus. S1; on ne peut être fort attentif à plusieurs choses en même temps. C1.

[885] § 86 C1.

[886] alia,] illo S1; illo obiecto S2; les autres, C1.

[887] tam integre disposita quam esse possent,] prorsus accurate ad actionem illam recipiendam disposita, S1, 97–98; autant disposés à en recevoir l'action qu'ils pourraient être, C1.

[888] ea disponat.] illam dispositionem S1, 98; les y disposer entièrement. C1.

[889] § 87 C1.

[890] 50.^{a} figura, quae parum est magis accurata, et inter alias partes oculi] figura 50: et inter eundem dispositum ad videndum obiectum paulo propius situm, uti ex figura 51 apparet: quae non in eo tantum consistit, quod humor crystallinus est paulo incurvatior, aliaeque oculi partes S1, 98–99; la 50. figure p. 65 et le même œil, disposé à en regarder un plus proche, comme il est en cette 51 qui consiste, non seulement en ce que l'humeur cristalline est un peu plus voûtée, et les autres C1, 87.

[891] in hac ultima] in posteriori S1, 99; en cette dernière C1.

[892] FF] H S1; H C1.

[893] pars, sive locus] pars S1; l'endroit C1.

[894] magis curvatus,] incurvatior, S1; plus voûté ou courbé; C1.

[895] in tubum 4 et a puncto C quod egrediuntur illi qui tendunt in tubum B.] in tubulum 2, ex puncto b in tubum 4, et ex puncto c in tubum 6. S1, 99–100; vers le tuyau 2; du point b, que sortent ceux qui tendent vers le tuyau 4; et du point c, que sortent ceux qui tendent vers le tuyau 6. C1.

[896] § C1.

[897] FF] H S1, 100; H C1.

[898] dispositionem corporis oculi,] toti oculo aliam dispositionem inducendam: S1; toute la disposition du corps de l'œil, C1, 88.

[899]Considera deinde tubos illos 2 4 6 posse eo magis aperiri per actionem obiecti ABC quo oculus magis dispositus est illud [93] respicere. Si enim radii qui cadunt in punctum[900] veniant verbi gr. omnes a puncto 6[901] prout faciunt, dum oculus fixe illud respicit, evidens est actionem illorum debere fortius trahere filum parvum 34[902] quam si venirent partim a puncto A partim a puncto B et partim a puncto C,[903] prout faciunt dum oculus parum alio modo est dispositus, quia[904] cum tunc eorum actiones[905,ccxxviii] non sint ita similes et unitae non possunt esse omnia[906] aeque fortes, imo saepe se mutuo impediunt, quod tamen non habet locum nisi in obiectis, quorum lineamenta non sunt nimis similia, neque nimis confusa, sicuti etiam non sunt praeter illa unde[907,ccxxix] oculus bene possit distinguere distantiam.[908] [909]Considera ulterius glandem FF[910] multo facilius posse moveri in partem in qua inclinando se, disponet oculum, ut recipiat actionem obiecti, quae omnium fortius aget in ipsum, et quidem distinctius, quam agat versus illos, versus quos[911] posset contrarium facere: prout verbi gratia [94] in hac 50.ᵃ figura,[912] multo minores vires esse necessarias ad incitandam ipsam ut inclinet se paulo magis versus anterius[913] quam requirantur ut retrocedat,[914] quia retrocedendo redderet oculum minus dispositum, quam sit ut reciperet actionem obiecti ABC[915,ccxxx] et sic esset in causa cur tubi parvi 2 4 6 illa actione essent minus aperti, et quod spiritus qui egrediuntur ex punctis A B C

---

[899] § 88 C1.

[900] punctum] punctum 3 S1, 101; le point 3 C1.

[901] 6] B S1; B C1.

[902] filum parvum 34] filamenta 3, 4, S1; le petit filet 3 4 C1.

[903] a puncto A partim a puncto B et partim a puncto C,] ab A, partim a B et partim a C. S1; du point A, partie de B, et partie de C, C1.

[904] quia] Huius rei ratio est, quia S1; à cause que C1.

[905] cum tunc eorum actiones] ipsorum actiones S1; ipsorum actiones tum temporis S2; pour lors leurs actions C1.

[906] omnia] prorsus S1; du tout C1.

[907] unde] quorum S1; dont C1.

[908] distantiam.] distantiam S1; la distance, et discerner les parties, ainsi que j'ai remarqué en la Dioptrique. C1.

[909] § 89 C1.

[910] FF] H S1; H C1.

[911] ut recipiat actionem obiecti, quae omnium fortius aget in ipsum, et quidem distinctius, quam agat versus illos [oculos], versus quos] ad recipiendam actionem obiecti, quod omnium efficacissime, et distinctissime in ipsum agit, quam ad eas partes, ubi S1; à recevoir plus distinctement qu'il ne fait l'action de l'objet qui agit le plus fort de tous contre lui, que vers ceux [côtés] où C1.

[912] figura,] figura S1; figure, où l'œil est disposé à regarder un objet éloigné, C1.

[913] versus anterius] antrorsum S1; en avant qu'elle n'est C1.

[914] requirantur ut retrocedat,] ut se recipiat retrorsum. S1; pour faire qu'elle se retire plus en arrière; C1.

[915] ABC] A, B, C S1; ABC, que l'on suppose être proche, et agir le plus fort de tous contre lui C1, 89.

caarent[916,ccxxxi] minus libere in istis tubis, ut si econtra approximaret faceret[917] ut oculus, disponens se ut illam actionem melius reciperet,[918] tubi parvi 2 4 6 aperirentur magis, et consequenter spiritus qui egrediuntur ex punctis A B C liberius ad illos fluerent. Ita ut statim[919] atque minimum quid incepisset moveri, cursus istorum spirituum eam statum[920,ccxxxii] auferret,[921] nec ipsi daret moram sistendi ⟦se⟧ se donec esset omnino disposita in modo quo vides illam sic dispositam in figura 51.ª et quod oculus [95] fixe in ABC[922] respiceret, [923]ita ut dicere solum refert causam, quae incipiat eam sic movendi, quae ordinarie non est alia quam vis ipsius obiecti, quae agens contra organum alicuius sensus auget introitum aliquorum ex parvis tubis, qui sunt in concava superficie cerebri, erga quos cum spiritus incipiant statim convolare, secum attrahunt glandem istam.[924] Sed posito quod isti tubi iam aliunde fuissent aeque aut magis aperti, quam obiectum illud illos aperiat, cogitandum est, quod partes parvae spirituum quae currant per poros eius,[925,ccxxxiii] cum sint inaequales, eam huc et illuc valde prompte minori tempore quam ictu oculi in omnes partes trudunt, absque quod remaneat quieta per unum momentum,[926] et si fiat statim ut illam trudunt versus unam partem versus quam ipsi non sit facile se inclinare,[927] eorum actio quae ex se non admodum vehemens, fere nullum effectum habere potest: sed e contra statim atque minimum [96] quid eam trudunt versus partem versus quam est iam omnino determinata, statim eo se inclinat et consequenter disponit organum sensus ut recipiat actionem sui obiecti quam fieri potest[928] perfectissime.

[929]Absolvamus nunc spiritus ad nervos usque ducere, videamusque motus qui ex iis dependent. Si enim tubi parvi partes[ccxxxiv] internae cerebri[930] non sint omnino magis aperti, nec aliqui aliter, quam alii, et consequenter si illi spiritus non habeant pressionem[931] alicuius ideae particularis diffundentur indifferenter[932] in omnes

---

[916] caarent] profluerent S1, 102; couleraient C1.

[917] faceret] effectum plane contrarium produceret S1; elle ferait tout au contraire C1.

[918] ut illam actionem melius reciperet,] ad actionem illam S1; mieux à recevoir cette action, C1.

[919] statim] simulac glandula S1; sitôt que la glande C1.

[920] statum] eodem momento S1; tout aussitôt C1.

[921] auferret,] abriperet transferretve S1; emporterait C1.

[922] ABC] A, B, C. S1; cet objet proche ABC. C1.

[923] § 90 C1, 90.

[924] glandem istam.] glandulam quadantenus S1; cette glande, et la font incliner vers ce côté-là. C1.

[925] eius [*i.e.:* glandulae],] eorum [*i.e.:* tubulorum] S1; ses [*i.e.:* de la glande] C1.

[926] absque quod remaneat quieta per unum momentum,] nullam requiem illi, vel per momentum concedentes. S1; sans la laisser jamais en repos un seul moment; C1.

[927] versus unam partem versus quam ipsi non sit facile se inclinare,] eo, quo se facile inclinare nequeat, S1, 103; vers un côté vers lequel il ne lui soit pas aisé de s'incliner, C1.

[928] potest] potest S1; possible, ainsi que je viens d'expliquer. C1, 91.

[929] § 32 S1; § 91 C1.

[930] partes internae cerebri] in intima cerebri parte S1, 104; la superficie intérieure du cerveau C1.

[931] pressionem] impressionem S1; en eux l'impression C1.

[932] indifferenter] temere atque indifferenter S1; indifféremment C1.

partes et transeunt poros qui sunt versus B in eos qui sunt versus C, unde subtiliores eorum partes omnino extra[933] cerebrum effluent per poros pellis parvae, qui eam circumvolvunt;[934,ccxxxv] deinde reliquum ubi ceperit cursum versus D[935] se ipsum in nervis et musculis recipiet absque eo quod causet effectum aliquem particularem, et quia in omnibus[936,ccxxxvi] aequaliter distribuitur. [937]Si autem aliqui ex istis tubis sint qui magis aut minus sint aperti[ccxxxvii]

Notes

i.	The underlining is just for emphasis.
ii.	Numeral absent in Schuyl's editions, probably because it is not part of an actual series of numerals; on the varying presence of numerals in the editions of *L'homme*, see Meschini 2016 (not considering the present case).
iii.	Probably referring to the two preceding changes.
iv.	Or: 'etiam'.
v.	The internal correction in Clerselier's edition seems to be of a typo.
vi.	The meaning of the period is different in Clerselier's and in the other versions; the *superscriptum* matches Clerselier's version (to which it might reveal an access, or to a further manuscript); "NB" might refer to the whole period.
vii.	Referring to "....demiarum."
viii.	The omission was certainly due to a difficulty in reading a manuscript. See Sect. 1.7.
ix.	Referring to "⟦carnem cordis⟧."
x.	This internal reference, absent in Schuyl's edition, is discussed in Meschini 2015. It may refer to the previously mentioned phenomena of fermentation, or to some explanation given in the missing chapters in between *Le monde* and *L'homme*.
xi.	Unclear reference.
xii.	Read: 'ad minimum'.
xiii.	Misspelled deleted contraction, for 'habent'.
xiv.	No new paragraphs in Schuyl and Clerselier.
xv.	Separator "/" added in the manuscript.
xvi.	"conatibus" is above "concutionibus." The *superscriptum* seems to be a variant unique to ATH 1444, and revealing the access to a further manuscript; the underlining just points to the subject of the juxtaposition.
xvii.	See Meschini 2015. It may refer to the differentiation of solids and liquids treated in chapter 3 of Descartes's *Le monde*.
xviii.	Mistaking of 'ces' and 'ses' by Schuyl.
xix.	Read: 'sit'.
xx.	Read: 'diversimode'.
xxi.	This lemma, "per transennam," reveals that De Raey was certainly not the author of the present translation, as he would have rendered it with 'obiter', as discussed in Sect. 1.7.
xxii.	The variant between the two editions by Clerselier seems to be a correction in transcription, or of a typo.

---

[933] omnino extra] ex S1; tout à fait hors S1.

[934] qui eam circumvolvunt;] quae eam involvit, S1; qui l'enveloppe; C1.

[935] deinde reliquum ubi ceperit cursum] residuae vero deinde spirituum partes profluentes S1; puis le surplus prenant son cours C1.

[936] in omnibus] omnibus S1; in omnes S2; en tous C1, 92.

[937] § C1.

xxiii.	No new paragraphs in Schuyl and Clerselier.
xxiv.	Read: 'transferendi'.
xxv.	The underlining might signify that the translation might contain a mistake (as in Schuyl's version), as 'recte' would be preferable (and matching Clerselier's version). This can point to the access to a further manuscript, or to their editions.
xxvi.	Referring to "<u>rectae</u>."
xxvii.	Referring to "<u>saliunt</u>," with a mutual underlining.
xxviii.	The marginal addition reveals the access to a French text different from Clerselier's version.
xxix.	Read: 'adhuc'. Probably a mistake in transcription.
xxx.	Read: 'simulac'.
xxxi.	Deleted beginning of a new paragraph. No new paragraphs in Schuyl and Clerselier.
xxxii.	Read: 'machinamentis'.
xxxiii.	Referring to "<u>eas</u>," with a mutual underlining. Between "commo" and "venti." The marginal addition might reveal the access to a further manuscript (and matching Clerselier's version), or might have been based on the preceding text, where machines are mentioned.
xxxiv.	The underlining signals a mistake in translation, as "speculae" is certainly used to translate 'regards' intended as lookouts rather than manholes or similar hydraulic structures. It might indicate the access to a further manuscript, or to Schuyl's or Clerselier's versions. The variant between the two editions by Clerselier seems to be a correction in transcription, or an editorial clarification.
xxxv.	Referring to "speculae."
xxxvi.	Read: 'aquae'.
xxxvii.	The underlining emphasizes the use of a French expression.
xxxviii.	Read: 'intra'.
xxxix.	Read: 'recondat'.
xl.	Deleted beginning of a new paragraph. No new paragraphs in Schuyl and Clerselier.
xli.	Unclear reference.
xlii.	Referring to "<u>tunica</u>," with a mutual underlining. Between "am" and "plo." The marginal addition reveals the access to a French text.
xliii.	For De Raey, "tubulos" (as in S1, 19) had to be rendered with 'filamenta', as discussed in Sect. 1.4.2. This occurrence was checked by Van Gutschoven.
xliv.	The reason for the underlining is unclear; it might be pointing to the characterization of nerves as double structures, viz. as consisting of two tunics or skins—a feature addressed by De Raey, as discussed in Sects. 1.4.2 and 1.7.
xlv.	Referring to "<u>tenui</u>."
xlvi.	Referring to "<u>inquam tubulus</u>." Between "tuni" and "cis."
xlvii.	Read: 'in quam'.
xlviii.	The underlining probably serves to point to the constitution of nerves as composed by multiple tubes (commented on by De Raey, as discussed in Sects. 1.4.2 and 1.7), or to the mistake in transcription of "<u>inquam</u>."
xlix.	According to Schuyl, this denotes the same structure intended by Descartes as the internal tubes contained in a nerve; though it in fact denotes the filaments composing the internal structure of the brain, and of the nerves, as discussed in Sect. 1.4.2.
l.	"isti" is above "[[ipsi]]." The correction might reveal the access to a further manuscript, or to Schuyl's or Clerselier's versions.
li.	For De Raey, "isti tubuli" (as in S1, 21) had to be rendered with 'ista filamenta', as discussed in Sect. 1.4.2. This occurrence was checked by Van Gutschoven.
lii.	The explanation of muscular movement, from here onwards, is a bit clearer in Clerselier's version, as mentioned in Sect. 1.7.
liii.	Between "C" and "G." Having been deleted, it might be referring to "HF," which is on the following page.
liv.	Referring to "<u>H [[F]]I</u>," with a mutual underlining.

lv.	Referring to "⟦F⟧I," with a mutual underlining. Between "ni" and "tuntur."
lvi.	Mistaking of 'passer' and 'pousser'.
lvii.	Read: 'post quae', or 'postque'. The underlining might refer to the mistake in transcription.
lviii.	Referring to "postquae."
lix.	The reason for the underlining is unclear. It might be due to a possible comparison with Schuyl's version, which is less literal, albeit using a similar verb.
lx.	Referring to "orsi."
lxi.	The underlining emphasizes the correction.
lxii.	Referring to "HF⟦E⟧I."
lxiii.	"⟦a⟧Anatomia" in the manuscript.
lxiv.	Mistaking of 'A' and 'et'.
lxv.	The rest of the page (c. four lines) is blank. No text is missing.
lxvi.	The underlining can point to the unclarity in using "exspirat," as, even if the period concerns respiration, what is at stake is the filling of muscles by spirits; alternatively, it points to its being a variant with respect to Schuyl's and Clerselier's versions (and to an access to them, or to a manuscript).
lxvii.	Reference, apparently, to "CG," even if not exactly on the same line.
lxviii.	Underlining for emphasis, probably given the relevance of the explanation of the process at stake, and its having been plagiarized by Regius.
lxix.	Read: 'egredi'.
lxx.	The underlining might refer to the importance of the verb in the description of the connection between the internal structure of the brain and the external nerves, as discussed in Sects. 1.4.2 and 1.7.
lxxi.	The variant between the two editions by Clerselier seems to be an editorial clarification.
lxxii.	Title only in ATH 1444.
lxxiii.	Read: 'suam'.
lxxiv.	"pellicul⟦{…}⟧ae" is above "cutes." The addition might indicate the access to a further manuscript, or to Schuyl's version; the underlining just points to the subject of the juxtaposition.
lxxv.	Read: 'inter ea'. Most probably a mistake in transcription.
lxxvi.	The sheets have been misplaced in the manuscript.
lxxvii.	The underlining points to a mistake in translation, as the sentence should be an infinitive.
lxxviii.	Referring to "sunt."
lxxix.	Title only in ATH 1444.
lxxx.	See Meschini 2015. Descartes does not provide a discussion of such particles in his *Le monde*, therefore, the reference might be to the missing chapters in between *Le monde* and *L'homme*. Nonetheless, he provides a discussion of them in his *Météores*, discourses 2 and 3.
lxxxi.	Referring to "tenerrimas." The symbol, of difficult rendering, seems to be repeated under "tenerrimas," though it might also be that both the symbol and the underlining have undergone a deletion.
lxxxii.	The reason for the underlining is unclear: probably, it was deleted. Alternatively, it might point to a variant with respect to Schuyl's version, though the two adjectives at stake are compatible with each other.
lxxxiii.	Referring to "⟦mutetur vis saporis⟧." Between "tem" and "peramentum." The underlining points to the deleted sentence. The marginal text partially matches Schuyl's version, while the deleted text matches Clerselier's version—so that the change might point to the access to Schuyl's version, or to a further manuscript.
lxxxiv.	Title only in ATH 1444.
lxxxv.	For a discussion of this variant see Sect. 1.7.
lxxxvi.	The deleted text matches Clerselier's version (so that it might point to the access to a further manuscript); the underlining, placed under the deleted text, might either have been for emphasis of it, and/or related to the following, interlinear "NB," as it starts after it.

lxxxvii.	"NB" is above "machina," referring either to the period—and in this case it would match Clerselier's version (to which it can reveal the access, or to a further manuscript)—or to the underlining below the deleted text.
lxxxviii.	The underlining might indicate a variant with respect to Schuyl's version, and reveal the access to it or to a further manuscript.
lxxxix.	Referring to "egreditur."
xc.	See Meschini 2015. The reference might be to the missing chapters in between *Le monde* and *L'homme*; there are no variants with respect to Schuyl's and Clerselier's versions.
xci.	Title only in ATH 1444.
xcii.	The deleted conjunction matches Clerselier's text, and might indicate the access to a further manuscript, or to Schuyl's version.
xciii.	The underlining might have been just for emphasis.
xciv.	For a discussion of this variant see Sect. 1.5.
xcv.	Read: 'ad minimum'.
xcvi.	Title only in ATH 1444.
xcvii.	"[e]Elementi" in the manuscript.
xcviii.	Mistaking of 'A, B, H' and 'ab H' in Schuyl's version, probably a typo.
xcix.	Unclear reference.
c.	The order of sheets has been misplaced in the manuscript.
ci.	The rest of the page (c. 16 lines) is blank.
cii.	Blank page.
ciii.	The underlining indicates a variant with respect to Schuyl's and Clerselier's version (and its access to them, or to a further manuscript), or a difficulty in making sense of the word.
civ.	Referring to "compassam." Between "compas" and "sam."
cv.	Read: 'congregantur'.
cvi.	Mistaking of 'sic' and 'sit' in Schuyl's version, probably a typo.
cvii.	Mistaking of 'et A' and 'et c.'.
cviii.	Mistaking of 'et A' and 'et c.'.
cix.	Unclear reference.
cx.	Apparently missing corresponding word. It might have been aimed at the insertion of a correction, as the passage contains a slight variant with respect to Clerselier's version (to which it might point the access, or to a further manuscript).
cxi.	Referring to the mentioned missing word.
cxii.	Read: 'alios'.
cxiii.	The underlining can indicate both a variant with respect to Schuyl's version (and the access to it), or to the philosophical significance of the verb.
cxiv.	Referring to "vadant."
cxv.	Read: 'fit'.
cxvi.	Mistake in translating 'de la mode'.
cxvii.	The omission was probably due to a difficulty in reading a manuscript. See Sect. 1.7.
cxviii.	Mistake in translating 'viennent'.
cxix.	Mistaking of 'ces' and 'ses', either in ATH 1444 and Schuyl's version, or in Clerselier's version.
cxx.	Mistaking of 'donc' and 'dont', either in ATH 1444 and Schuyl's version, or in Clerselier's version.
cxxi.	Mistake in translating 'leurs' in Schuyl's version.
cxxii.	Mistake in translating 'leur' in Schuyl's version.
cxxiii.	Read: 'ratione'.
cxxiv.	Mistaking of 'ces' and 'ses', either in ATH 1444 and Schuyl's version, or in Clerselier's version.
cxxv.	ATH 1444 and the main text of Schuyl's edition seem to be subjected of a transcription error and a typo; this might reveal the access of the copyist of ATH 1444 to Schuyl's version, or vice-versa.

Notes

cxxvi.	Mistaking, in transcription, of 'si' and 's.' or 'sc.'.
cxxvii.	Underlining for emphasis.
cxxviii.	Referring to "N." Between "retrac" and "tus."
cxxix.	The text should read: 'lineae, quorum vi actiones obiectorum distantium transeunt per sensus'; in which case, it agrees with Clerselier's version.
cxxx.	Read: 'illas'.
cxxxi.	Or: 'ordinaria'.
cxxxii.	Read: 'ceram'.
cxxxiii.	The variant in Schuyl's second edition was probably just typographic.
cxxxiv.	Probable mistake in transcription.
cxxxv.	The lemmas concerning the parts of the organs show different translations in the present paragraph.
cxxxvi.	Apparent error in translation or transcription.
cxxxvii.	Schuyl's version points to the pores of nerves, while in fact the reference is more properly to the intervals or *mailles* between the filaments of the brain, as discussed in Sect. 1.4.2.
cxxxviii.	Read: 'motus', or 'modos'.
cxxxix.	Read: 'anatomici'.
cxl.	Probable mistake in transcription by omission.
cxli.	This numeral, showing no variants with respect to the other versions, is—in Clerselier's version—the first of a series commented on in Meschini 2011.
cxlii.	Misspelled deleted contraction, for 'proprii'.
cxliii.	As to the variants concerning the treatment of the passions, provided in this paragraph, see Sect. 1.7.
cxliv.	Mistaking, in transcription, of 'aliquando' and 'aliquomodo'.
cxlv.	Mistaking of 'embrasser' and 'embraser'.
cxlvi.	This lemma—corresponding to 'd'autant'—shows different translations in the present paragraph.
cxlvii.	Mistaking of 'embrasser' and 'embraser'.
cxlviii.	Mistaking of 'resserrer' and 'réserver'.
cxlix.	Mistaking of 'embrasser' and 'embraser'.
cl.	Probable mistaking of 'regarder' and 'regorger'.
cli.	The underlining might point to the (transcription) variant with respect to the versions of Schuyl and Clerselier (to which it can reveal the access, or to a further manuscript), or just to the difficulty in making sense of the sentence.
clii.	Referring to "vivorum."
cliii.	Read: 'fumus'.
cliv.	This variant reveals a series of numerals present, in its second and third occurrences, only in Clerselier's version, commented on in Meschini 2011 (arguing that it was an author's variant, as in Clerselier's version the numeral is awkwardly at the beginning of a new part of the treatise).
clv.	This variant reveals that De Raey was certainly not the author of the present translation, as discussed in Sect. 1.7.
clvi.	Mistaking of 'donc' and 'dont'.
clvii.	This omission, probably due to a difficulty in making sense of the text rather than of deciphering it, reveals that De Raey, who criticized Schuyl's translation of it, was certainly not the author of the present translation, as discussed in Sect. 1.7.
clviii.	Referring to "........".
clix.	In Clerselier's editions "48" and "49" are on two sides of figure M (prepared by La Forge: C1, 64), while "48" designates also an independent figure (prepared by Van Gutschoven: C1, 63 and 65); in Schuyl's version figures 48 and 49 are two independent figures, and numbered as figures 31 and 34. The variant between the two editions by Clerselier seems to be an editorial clarification.
clx.	Mistaking of 'donc' and 'dont'.

clxi.	This variant reveals that De Raey was certainly not the author of the present translation, as discussed in Sect. 1.7.
clxii.	Plain mistake in translating 'vaisseaux'.
clxiii.	Third numeral of the series discussed in Meschini 2011.
clxiv.	Numeral absent in Clerselier's version, probably because it is not part of an actual series of numerals.
clxv.	Read: 'vivas'. Most probably a mistake in transcription, as the masculine does not agree with 'partes' (viz. "suarum partium").
clxvi.	Read: 'superfluae'. Most probably a mistake in transcription, as it does not agree with 'partes'.
clxvii.	Read: 'conductos'. Most probably a mistake in transcription, as it does not agree with "qui," and the subsequent mentioning of "I," evidently one of the canals. The use of the masculine in the cases of "vivos," "superflui," and "agitatos" might be explained as an attempt to make them agree with "qui"—not identified with the canals—rather than with 'partes'.
clxviii.	Read: 'agitatae'. Most probably a mistake in transcription, as it does not agree with 'partes'.
clxix.	Read: 'adhuc'. Certainly a mistake in transcription.
clxx.	The variant between the two editions by Clerselier seems to be a correction in transcription, or of a typo.
clxxi.	As to the variants concerning the treatment of the pineal gland, see Sect. 1.7.
clxxii.	Mistaking of 'deux' and 'd'eux' in Schuyl's version.
clxxiii.	Read: 'affixa'.
clxxiv.	Read: 'repraesentat'.
clxxv.	Read: 'exspansam'.
clxxvi.	Read: 'inflet'.
clxxvii.	The meaning of the sentence is different in ATH 1444, probably due to a mistaking of 'le' and 'se', as in the former case Clerselier's version would match ATH 1444.
clxxviii.	According to Schuyl, this denotes the same structure intended by Descartes as the internal tubes contained in a nerve; though it in fact denotes the openings or interstices between the filaments composing the internal structure of the brain, as discussed in Sect. 1.4.2. For Schuyl, this applies to the structures labelled 'tuyaux' which he translated with 'tubuli' and 'tubi' at page 80 of his edition (7 occurrences, up to "canales 2 4 6" in ATH 1444, 77).
clxxix.	The internal correction in Clerselier's edition seems just to be of a typo.
clxxx.	No new paragraphs in Schuyl and Clerselier.
clxxxi.	The underlining points to some ambiguity certainly present in the original French, such as that one vitiating a full comprehension of the sentence in Clerselier's version, where one should read, instead of 'celle', 'celles', which agrees with "tracent" (C1, 72): viz. the ways in which spirits move out of the pores or points of the gland, so that such ways enable the tracing of a figure on the surface of the gland itself. By reading 'celle' (viz. figure), in turn, one would think that the figure itself, according to which the spirits move out of the gland, traces a figure on the surface of the gland. Schuyl provides a clearer version, i.e. he overtly mentions such ways as tracing a figure on the gland. According to his version, however, it is unclear to which kinds of ways the text refers: being determined by the spirits themselves, these should be the ways in which the points of the gland are opened by the spirits—as the opening of the pores in the internal substance of the brain are determined by the movement of the filaments contained in the nerves. In the other versions, the reference is more clearly just to the ways in which spirits move.
clxxxii.	Read: 'quas'.
clxxxiii.	The underlining points out to the fact that this is an internal quotation.
clxxxiv.	The underlining might be for emphasis, or to point out a minor variant.
clxxxv.	Referring to "<u>embrionis</u>."

clxxxvi.	This variant reveals that De Raey was certainly not the author of the present translation, as discussed in Sect. 1.7.
clxxxvii.	The variant between the two editions by Clerselier seems to be a correction in transcription, or of a typo.
clxxxviii.	Read: 'retractata'.
clxxxix.	Read: 'delentur'.
cxc.	Mistake in translating 'dont'.
cxci.	Read: 'illae'.
cxcii.	Mistaking, in transcription, of 'uno' and 'imo'.
cxciii.	Read: 'illum'.
cxciv.	Mistake in transcription.
cxcv.	Read: 'aliquas'.
cxcvi.	The variant between the two editions by Clerselier is the correction of a typo.
cxcvii.	Read: 'parum'.
cxcviii.	According to Schuyl, this denotes the same structure intended by Descartes as the internal tubes contained in a nerve; though it in fact denotes the openings or interstices between the filaments composing the internal structure of the brain, as discussed in Sect. 1.4.2. For Schuyl, this applies to the structures labelled 'tuyaux' which he translated with 'tubuli' at page 89 of his edition (3 occurrences in his edition, up to the next 'canales parvos').
cxcix.	Probably a mistake in translation, due to the mistaking of 'les' and 'le', as in the former case Clerselier's version would match ATH 1444.
cc.	In this case, as discussed in Sect. 1.7, Schuyl changed his text, in his second edition, in the light of a version matching those of ATH 1444 and of Clerselier. The text of ATH 1444 is rather inconsistent; one should read 'haec ipsa' or 'hoc ipsum'.
cci.	Read: 'aliquos'.
ccii.	In this case, as discussed in Sect. 1.7, Schuyl changed his text, in his second edition, in the light of a version matching those of ATH 1444 and of Clerselier.
cciii.	Read: 'aliquos alios'.
cciv.	In this case, as discussed in Sect. 1.7, Schuyl changed his text, in his second edition, in the light of a version matching those of ATH 1444 and of Clerselier.
ccv.	In this case, as discussed in Sect. 1.7, Schuyl changed his text, in his second edition, in the light of a version not matching the other ones.
ccvi.	Read: 'motus'.
ccvii.	Apparently missing corresponding word.
ccviii.	Probably a mistake in transcription; the previous insertion of a reference mark might have been due to the attempt at making sense of it, by inserting a corresponding—and missing—verb.
ccix.	The addition might reveal the access to a further manuscript.
ccx.	Read: 'sit'.
ccxi.	Read: 'sit'.
ccxii.	"in" is above "⟦ha⟧."
ccxiii.	The addition, if not just a correction of an error in transcription (as suggested by the deletion), might reveal the access to Clerselier's version, or to a further manuscript.
ccxiv.	Mistaking of 'ces' and 'ses', either in ATH 1444 and Schuyl's version, or in Clerselier's version.
ccxv.	In this case, as discussed in Sect. 1.7, Schuyl changed his text, in his second edition, in the light of a version matching that of Clerselier.
ccxvi.	In this case, as discussed in Sect. 1.7, Schuyl changed his text, in his second edition, in the light of a version matching those of ATH 1444 and of Clerselier.
ccxvii.	Mistaking, in transcription, of 'aliquando' and 'aliquo'.
ccxviii.	Read: 'fere'.
ccxix.	Read: 'hoc'.
ccxx.	Read: 'aa' for 'anima'.

ccxxi.	According to Schuyl, this denotes the same structure intended by Descartes as the internal tubes contained in a nerve; though it in fact denotes the openings or interstices between the filaments composing the internal structure of the brain, as discussed in Sect. 1.4.2. For Schuyl, this applies to the structures labelled 'tuyaux' which he translated with 'tubi' at page 97 of his edition.
ccxxii.	Mistaking of 'donc' and 'dont'. In this case, as discussed in Sect. 1.7, Schuyl changed his text, in his second edition, in the light of a version matching that of Clerselier.
ccxxiii.	The variant in Schuyl's second edition seems to be a translation variant.
ccxxiv.	In this case, as discussed in Sect. 1.7, Schuyl changed his text, in his second edition, in the light of a version matching those of ATH 1444 and of Clerselier.
ccxxv.	In this case, as discussed in Sect. 1.7, Schuyl changed his text, in his second edition, in the light of a version matching that of Clerselier.
ccxxvi.	The variant in Schuyl's second edition seems to be a translation variant.
ccxxvii.	As discussed in Sect. 1.7, a part of the text is evidently missing: this includes a reference to another figure afterwards alluded to in the text (so that the omission was certainly a mistake), and makes the whole argument quite hard to understand, if not inconsistent (while its being at least grammatically consistent suggests that someone tried to correct the text).
ccxxviii.	In this case, as discussed in Sect. 1.7, Schuyl changed his text, in his second edition, in the light of a version matching those of ATH 1444 and of Clerselier.
ccxxix.	Mistaking of 'donc' and 'dont'.
ccxxx.	As to the variants concerning the treatment of the pineal gland, see Sect. 1.7.
ccxxxi.	Probably a misspelled contraction, for 'colarent'.
ccxxxii.	Read: 'statim'.
ccxxxiii.	Mistake in translation in Schuyl's version, probably due to a mistaking of 'les' and 'ses' and to the consequent difficulty in intending to which structure the pores belong.
ccxxxiv.	Read: 'partis'.
ccxxxv.	Apparent mistakes in translation both in ATH 1444 and in Schuyl's version, in both cases due to a difficulty in understanding for what *l'* (rendered with "eam") stands for. According to ATH 1444, the pores envelop the membrane, so that the verb is translated as a plural; according to Schuyl's version, the membrane envelops the internal part of the brain, mentioned earlier in the paragraph; alternatively, "eam" should be read as a typo for 'eum' in Schuyl's version, and meaning the brain as such (as it appears to be in Clerselier's version).
ccxxxvi.	In this case, as discussed in Sect. 1.7, Schuyl changed his text, in his second edition, in the light of a version matching that of Clerselier.
ccxxxvii.	Reference word for the text in the next page, which is missing.
ccxxxviii.	eius] earum S1; leur C1.
ccxxxix.	ei circumstantium] circumstant S1; les environnent C1, 6–7.
ccxl.	Read: 'constans'. Otherwise, it could be seen to agree with "unoquoque": this could lead to further problems in interpreting Descartes's neuroanatomy.
ccxli.	Read: 'dividatur'.
ccxlii.	Read: 'eorum'.
ccxliii.	Read: 'quem'.
ccxliv.	Read: 'tenui'. Otherwise, it could be seen to agree with 'fila'.
ccxlv.	Read: 'situm'. Otherwise, it could be seen to agree with "manus" or "fila."
ccxlvi.	Read: 'qui'.
ccxlvii.	Read: 'dispositus'.
ccxlviii.	Read: 'ultimas plicas' or 'ultimam plicam'.
ccxlix.	Read: 'ipsis'.

Appendix

Figures

Fig. A.1 Schuyl's representation of the constitution of the brain, nerves, and muscles (*Source* Descartes 1662, 20. University of California, Biomed History and Special Collections, signature: WZ 250 D453d 1662. Public domain)

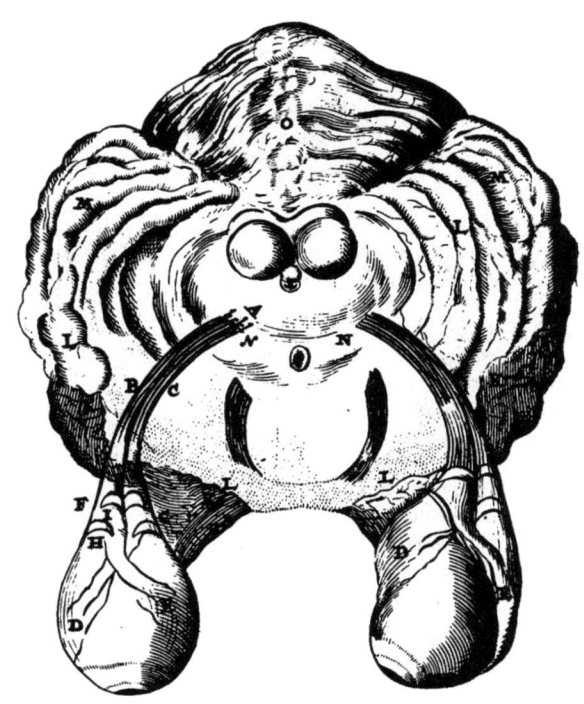

Fig. A.2 Van Gutschoven's representation of the nerve-muscle system (*Source* Descartes 1664b, 16. University of California, Biomed History and Special Collections, signature: WZ 250 D453h 1664. Public domain)

Fig. A.3 Descartes's representation of the nerve-muscle system (in Clerselier's edition) (*Source* Descartes 1664b, 17. University of California, Biomed History and Special Collections, signature: WZ 250 D453h 1664. Public domain)

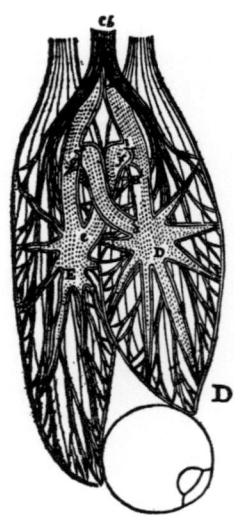

Appendix 113

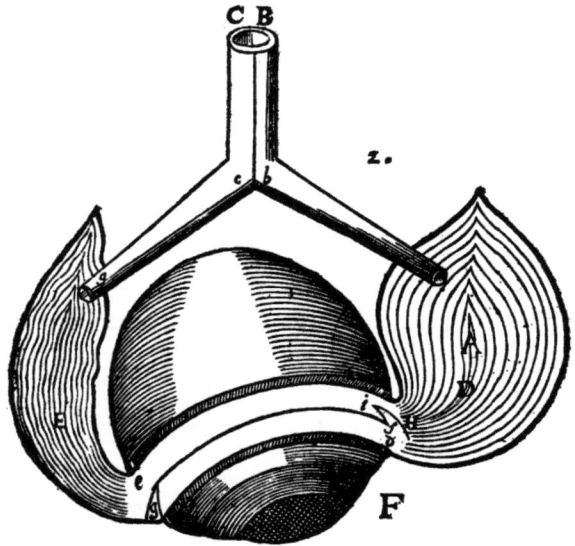

Fig. A.4 La Forge's representation of the nerve-muscle system (*Source* Descartes 1664b, 18. University of California, Biomed History and Special Collections, signature: WZ 250 D453h 1664. Public domain)

Fig. A.5 Descartes's representation of the nerve-muscle system (in Schuyl's edition) (*Source* Descartes 1662, 25. University of California, Biomed History and Special Collections, signature: WZ 250 D453d 1662. Public domain)

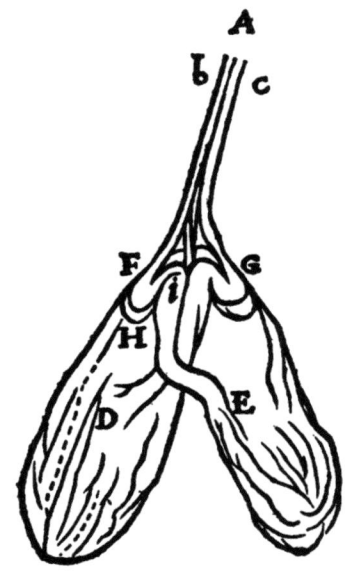

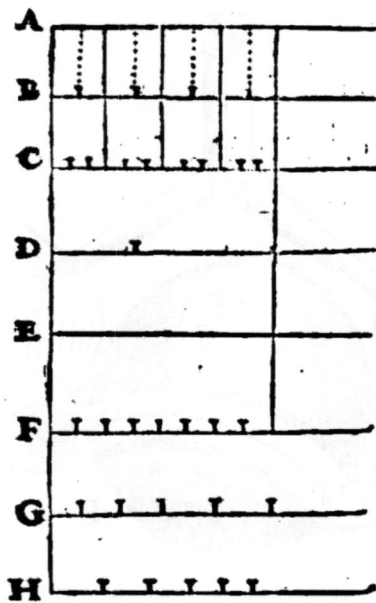

Fig. A.6 Descartes's representation of sounds (in Schuyl's edition) (*Source* Descartes 1662, 43. University of California, Biomed History and Special Collections, signature: WZ 250 D453d 1662. Public domain)

Fig. A.7 The representation of sounds in Clerselier's edition (*Source* Descartes 1664b, 36. University of California, Biomed History and Special Collections, signature: WZ 250 D453h 1664. Public domain)

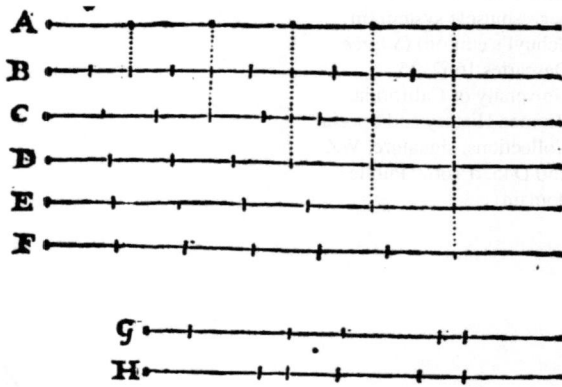

Appendix

Tables

Table A.1 Textual reliance of Huyberts on Descartes's *L'homme*

Huyberts's *Disputatio* (1652)	Clerselier's edition (1664)
I. Antequam aliquid nutrire corpora nostra possit, alterari, dissolvi, et in liquorem converti debet. Cui mutationi *cibi ex natura sua obnoxii sunt: **talis enim plerumque naturae sunt, ut sponte putrescere, dissolvi, et incalescere possint, haud aliter ac foenum recens antequam probe siccatum est, in horreis reconditum.** Valde autem tarda esset haec dissolutio (…) V. Attendenti autem porro ad ventriculi constitutionem manifestum erit in eum †**talem liquorem** influere posse, atque a crassiori sanguinis massa ibi separari. **Ob arterias** enim quas insignes habet multum sanguinis ad ipsum confluit, qui **necessario quoque valde calidus esse debet** et spirituosus, propterea quod viae per quas fluit ita constitutae sint, ut sanguis subtilior magisque motu ob naturae leges a reliqua massa per eas deflectere debeat: cuius rursus partes tenuiores magisque ad motum aptae, quales praecipuae sunt spiritus acres et oleaginei, per arteriarum tunicas earumque extremitates in ipsam ventriculi cavitatem transudant, ubi omnes ‡**cibos penetrando eorum partes haud aliter quam aqua communis calcis vivae aut aqua fortis metallorum partes disiungit, movet, et calefacit,** uno verbo fermentat. (…)	Premièrement ‡les viandes se digèrent dans l'estomac de cette machine, par la force de certaines liqueurs, qui, se glissant entre leurs parties, les séparent, les agitent, et les échauffent, ainsi que l'eau commune fait celles de la chaux vive, ou l'eau forte celles des métaux. Outre que †ces liqueurs étant apportées du cœur fort promptement par les artères, ainsi que je vous dirai ci-après, ne peuvent manquer d'être fort chaudes. Et même *les viandes sont telles pour l'ordinaire qu'elles se pourraient corrompre et échauffer toutes seules, ainsi que fait le foin nouveau dans la grange, quand on l'y serre avant qu'il soit sec. (…) **Lors que les liqueurs, que j'ai dit cl-dessus servir comme d'eau-forte dans son estomac, et y entrer sans cesse de toute la masse du sang par les extrémités des artères, n'y trouvent pas assez de viandes à dissoudre pour occuper toute leur force, elles la tournent contre l'estomac même, et agitant les petits filets de ses nerfs plus fort que de coutume, font mouvoir les parties du cerveau d'où ils viennent: ce qui sera cause que l'âme étant unie à cette machine concevra l'idée générale de la faim. Et ††si ces liqueurs sont disposées à employer plutôt leur action contre certaines viandes particulières que contre d'autres, ainsi que l'eau forte commune dissout plus aisément les métaux que la cire, elles agiront aussi d'une façon particulière contre les nerfs de l'estomac, laquelle sera cause que l'âme concevra pour lors l'appétit de manger de certaines viandes, plutôt que d'autres. (*Hic notari potest mira huius machinae conformatio, quod fames oriatur ex ieiunio; sanguis enim circulatione acrior fit; et ita liquor ex eo in stomachum veniens nervos magis vellicat; idque modo peculiari, si peculiaris sit constitutio sanguinis; unde pica mulierum*)[1]

(continued)

---

[1] Descartes 1664b, 2–3 and 55–56; italics by Clerselier.

Table A.1 (continued)

Huyberts's *Disputatio* (1652)	Clerselier's edition (1664)
VI. ****Quia autem perpetuo sanguis per ventriculi arterias fluit, perpetuo quoque liquor praedictus in eum instillat,** qui vires suas exercet in id quod ibi **invenit.** Si ergo contingat ventriculum vacuum esse, aut **id** quod in ventriculo est **non sufficere in quo eius** actio et **vires terminentur,** necessarium est ut **ipsum ventriculum eas convertat, cuius nervulos solito magis movendo famis sensus in nobis excitat.** (…) VIII. Nam, ††**si** consideremus **menstrua** solvendis corporibus destinata debere constare **mixtura et proportione partium appropriata proposito corpori solvendo, eorumque actiones prout constitutio eorum variat aut aliquid iis accedit variare,** idemque corpus vario modo ab iis affici posse, **quemadmodum videmus aquam fortem magis aptam esse ad** aes, ferrum, aliaque **metalla dissolvenda, quam ceram,** et aurum quod ab ea solvi non potest nisi aliquid salis ammoniaci in ea ante solutum fuerit, (…) IX. Et quoniam menstruum illud non solum vires suas exercet in cibos sed etiam in ipsum ventriculum (in qua actione famem consistere dixi) **hinc fit, ut, prout illud dispositum est, varie etiam in eum agat, varioque modo eius nervulos movendo varias famis species seu appetitus in nobis excitet.** Haud aliter ac videmus saporum multitudinem (…)[2]	

The use of bold indicates those parts of Huyberts's text that show textual agreement or similarity with Descartes's treatise; corresponding parts of the texts have been pinpointed by dedicated symbols: *, †, ‡, **, and ††

---

[2] Huyberts 1652, theses 1, 5, 6, 8, and 9.

Table A.2 Paragraph division of ATH 1444

ATH 1444, unnumbered paragraphs and page numbers	Schuyl's editions	Clerselier's 1664 edition
1 (1)	1	1
2 (1)	2	2
3 (2)	3	3
4 (4)	No new paragraph	4
5 (5)	4	6
6 (6)	No new paragraph	Unnumbered new paragraph
7 (7)	6	7
8 (7)	No new paragraph	No new paragraph
9 (8)	7	8
10 (9)	8	9
11 (10)	No new paragraph	10
12 (12)	9	12
13 (12)	10	13
14 (13)	No new paragraph	No new paragraph
15 (13)	No new paragraph	14
16 (14)	No new paragraph	Unnumbered new paragraph
17 (15)	No new paragraph	Unnumbered new paragraph
Deleted new paragraph divider (16)	No new paragraph	No new paragraph
Deleted new paragraph divider (18)	No new paragraph	No new paragraph
18 (18)	No new paragraph	17
19 (19)	No new paragraph	18
20 (21)	No new paragraph	20
21 (21)	No new paragraph	21
22 (25)	No new paragraph	23
23 (27)	11	24
24 (28)	12	25
25 (30)	13	26
26 (32)	No new paragraph	27
27 (33)	No new paragraph	28
28 (38)	14	32
29 (41)	15	34
30 (43)	16	35

(continued)

Table A.2 (continued)

ATH 1444, unnumbered paragraphs and page numbers	Schuyl's editions	Clerselier's 1664 edition
31 (45)	17	No new paragraph
32 (46)	18	37
33 (hypothetical, in a missing page before page 49)	19	40
34 (55)	20	47
35 (56)	No new paragraph	48
36 (57)	No new paragraph	Unnumbered new paragraph
37 (59)	No new paragraph	Unnumbered new paragraph
38 (63)	21	Unnumbered new paragraph
39 (63)	22	52
40 (64)	23	No new paragraph
41 (65)	24	55
42 (68)	27	57
43 (70)	No new paragraph	61
44 (70)	No new paragraph	62
45 (71)	29	Unnumbered new paragraph
46 (72)	No new paragraph	64
47 (72)	No new paragraph	Unnumbered new paragraph
48 (73)	No new paragraph	Unnumbered new paragraph
49 (75)	30	66
50 (77)	No new paragraph	No new paragraph
51 (77)	No new paragraph	69
52 (82)	No new paragraph	75
53 (89)	No new paragraph	84
54 (90)	No new paragraph	85
55 (96)	32	91

Bibliography

Handwritten Sources

Amsterdam, Stadsarchief, toegangsnummer 5001, inventarisnummer 41
Amsterdam, Stadsarchief, toegangsnummer 728, inventarisnummer 37
Amsterdam, Universiteitsbibliotheek, ms. K 137
Copenhagen, Det Kongelige Bibliotek, ms. Don. var. nr. 145 4to, *Annotata ad Principia philosophica Renati Des-Cartes, excepta in collegio, habito sub Johanne de Raei, inchoato die 1 Maii 1658, finito die 20 Decembris* = De Raey 1658
Groningen, Universiteitsbibliotheek, ms. GN108
Hamburg, Staats- und Universitätsbibliothek Carl von Ossietzky, Cod. phil. 323 W. 28, 1–231, *Dictata clarissimi atque acutissimi Domini Johannis de Raei (…) in Dissertationem de methodo Renati des Cartes (…). Dictata (…) in Principia philosophiae*, c. 1659–1661 = De Raey 1659–1661
Leiden, Erfgoed Leiden en Omstreken, toegangsnummer 1004, inventarisnummer 222
Leiden, Erfgoed Leiden en Omstreken, toegangsnummer 1004, inventarisnummer 23 Y
Leiden, Universiteitsbibliotheek, ms. ATH 1444, *Tractatus de homine a Cartesio*. URL = <http://hdl.handle.net/1887.1/item:3487408>. Accessed 23 March 2024 = ATH 1444
London, Wellcome Library, ms. 3415, 1–111 (independent numbering), *Dictata clarissimi doctissimique viri, D. Johannis de Raeij, ad Epitomen Institutionum medicarum viri celeberrimi Danielis Sennerti*, 1656 or 1661 = De Raey 1656/1661
Paris, Bibliothèque Nationale de France, MS latin 10352, *Epistolae Samuelis Sorbière ad illustres et eruditos viros scriptae (…). Accedunt illustrium et eruditorum virorum ad eundem Epistolae (…) Cura et opera Henrici Sorbière, auctoris filii, Parisiis*, 1673 = Sorbière 1673
The Hague, Koninklijke Bibliotheek, ms. KA 45
Uppsala, Universitetsbiblioteket, Waller Ms benl-00586
Utrecht, Het Utrechts Archief, toegangsnummer 34-4, inventarisnummer U048a004
Utrecht, Het Utrechts Archief, toegangsnummer 711, inventarisnummer 129
Utrecht, Het Utrechts Archief, toegangsnummer 711, inventarisnummer 99

Other Sources

Agostini 2009 = Agostini, Siegrid. 2009. *Claude Clerselier editore e traduttore di René Descartes*. Lecce: Conte Editore.

Alexandrescu 2012 = Alexandrescu, Vlad. 2012. "What Someone May Have Whispered in Elisabeth's Ear." In *Oxford Studies in Early Modern Philosophy. Volume VI*, ed. Daniel Garber and Donald Rutherford, 1–27. Oxford: Oxford University Press. https://doi.org/10.1093/acprof:oso/9780199659593.003.0001

Andreae 1653a = Andreae, Tobias. 1653. *Methodi Cartesianae assertio opposita Jacobi Revii praefatae methodi cartesianae Considerationi theologicae quam vocat*. Groningen: Typis Joannis Cöelleni.

Andreae 1653b = Andreae, Tobias. 1653. *Brevis replicatio reposita Brevi explicationi mentis humanae, sive animae rationalis D. Henrici Regii*. Amsterdam: Typis Ludovici Elzevirii.

Anonymous (Sallo?) 1665 = Anonymous (Denis de Sallo?). 1665. "*L'Homme* de René Descartes, avec un *Traité de la formation du foetus* du mesme autheur. A Paris." *Journal des sçavans*, 5 January 1665, 9–11.

Antoine-Mahut and Gaukroger 2016 = Antoine-Mahut, Delphine, and Gaukroger, Stephen (eds.). 2016. *Descartes' Treatise on Man and Its Reception*. Dordrecht: Springer. https://doi.org/10.1007/978-3-319-46989-8

Aucante 2006 = Aucante, Vincent. 2006. *La philosophie médicale de Descartes*. Paris: Presses Universitaires de France.

Baillet 1691 = Baillet, Adrien. 1691. *La Vie de M. Des-Cartes*. Paris: Chez Daniel Horthemels.

Bartholin 1663–1667 = Bartholin, Thomas. 1663–1667. *Epistolarum medicinalium a doctis vel ad doctos scriptarum, centuria I et II[–IV]*. Copenhagen: Typis M. Godicchenii.

Belgioioso 2005 = Belgioioso, Giulia. 2005. "Un faux de Clerselier." *Bulletin cartésien* XXXIII/*Archives de Philosophie* 68(1), 148–158.

Bitbol-Hespériès 2019 = Bitbol-Hespériès, Annie. 2019. "Des gravures de Descartes dans l'édition parisienne de *L'Homme* en 1664?" *Bulletin cartésien* XLVIII/*Archives de philosophie* 50(1), 157–165.

Boissière 1909 = Boissière, Gustave. 1909. *Urbain Chevreau (1613-1701). Sa vie – ses oeuvres*. Niort: G. Clouzot.

Borch 1983 = Borch, Ole. 1983. *Olai Borrichii Itinerarium 1660-1665. The Journal of the Danish Polyhistor Ole Borch*, ed. Henrik D. Schepelern. Copenhagen-London: The Danish Society of Language and Literature.

Bos 2004 = Bos, Erik-Jan. 2004. "La Lettre AT no. 585 réexaminée." *Bulletin cartésien* XXXV/*Archives de philosophie* 70(1), 9–13.

Bos 2017 = Bos, Erik-Jan. 2017. "Descartes and Regius on the Pineal Gland and Animal Spirits, and a Letter of Regius on the True Seat of the Soul." In *Descartes and Cartesianism: Essays in Honour of Desmond Clarke*, ed. Stephen Gaukroger and Catherine Wilson, 95–111. Oxford: Oxford University Press. https://doi.org/10.1093/acprof:oso/9780198779643.003.0006

Bos 2022 = Bos, Erik-Jan. 2022. "An Unknown Latin Manuscript Translation of Descartes' 'L'Homme'/Onbekende vertaling van René Descartes' 'L'homme' ontdekt in Leidse Bibliotheca Thysiana." *Leiden Special Collections Blog*, 18 August 2022. URL = <https://www.leidenspecialcollectionsblog.nl/articles/an-unknown-latin-manuscript-translation-of-descartes-lhomme>, and URL = <https://www.universiteitleiden.nl/in-de-media/2022/08/onbekende-vertaling-van-rene-descartes-lhomme-ontdektin-leidse-bibliotheca-thysiana>. Accessed 23 March 2024.

Breyne 1678 = Breyne, Jakob. 1678. *Exoticarum aliarumque minus cognitarium plantarum centuria prima*. Gdańsk: Imprimebat David Fridericus Rhetius.

Caps 2010 = Caps, Géraldine. 2010. *Les « médecins cartésiens ». Héritage et diffusion de la représentation mécaniste du corps humain (1646–1696)*. Hildesheim-Zürich-New York: Georg Olms Verlag.

Catalogus 1657 = s. a. 1657. *Catalogus librorum (…) Antonii Thysii*. Leiden: Ex typographia Philippi de Croy. Paris, Bibliothèque Nationale de France, signature Q 2317

Catalogus 1666 = s. a. 1666. *Catalogus insignium in omni facultate librorum bibliothecae (…) Antonii Thisii*. Leiden: Apud Philippum De Cröy. St. Petersburg, Rossijskaja nacional'naja biblioteka, signature 16.20.9.305

Bibliography 121

Catalogus 1668 = s. a. 1668. [*Catalogus bibliothecae Joan. Thysii*]. [Leiden]: s.n. Leiden, Universiteitsbibliotheek, signature THYSIA 258

Catalogus 1677 = s. a. 1677. *Catalogus bibliothecae* (…) *Joannis Thysii, institutae in perpetuam sui memoriam et usum posterorum*. Leiden: Ex typographia Arnoldi Doude. Leiden, Universiteitsbibliotheek, signature 1402 D 21:1

Catalogus 1723 = s. a. 1723. *Bibliotheca Schalbruchiana, sive catalogus exquisitissimorum rarissimorumque librorum* (…) *quos* (…) *collegit Joh. Theod. Schalbruch*. Amsterdam: Apud R. et G. Wetstenios. Ghent, Universiteitsbibliotheek, signature BW 83

Catalogus 1739 = s. a. 1739. *Catalogus bibliothecae* (…) *Johannis Thysii, institutae in perpetuam sui memoriam et usum posterorum*. Leiden: Ex typographia Joh. Wilhelmi De Groot.

Catalogus 1852 = s. a. 1852. *Catalogus librorum bibliothecae Thysianae, in Academia Lugduno-Batava*. Leiden: Ex typographeo J. G. La Lau.

Catalogus 1879 = Tiele, P. A. 1879. *Catalogus der bibliotheek van Joannes Thysius*. Leiden: E. J. Brill.

Chanut 1677 = Chanut, Pierre-Hector. 1677. *Memoires de ce qui s'est passé en Suede, et aux provinces voisines, depuis l'année 1645 jusques en l'année 1655*. Cologne [Bruxelles]: Chez Pierre Du Marteau [Eugène Henry Fricx].

Chevreau 1697 = Chevreau, Urbain. 1697. *Chevraeana*. Paris: Chez Florentin et Pierre Delaulne.

Clauberg 1652 = Clauberg, Johannes. 1652. *Defensio cartesiana, adversus Iacobum Revium* (…) *et Cyriacum Lentulum* (…)*: pars prior exoterica, in qua Renati Cartesii Dissertatio de methodo vindicatur, simul illustria cartesianae logicae et philosophiae specimina exhibentur*. Amsterdam: Apud Ludovicum Elzevirium.

Clauberg 1655 = Clauberg, Johannes. 1655. *Initiatio philosophi, sive Dubitatio cartesiana, ad metaphysicam certitudinem viam aperiens*. Leiden: Ex officina A. Wyngaerden.

Cook 2021 = Cook, Harold J. 2021. "Princess Elisabeth's Cautions and Descartes' Suppression of the *Traité de l'Homme*." *Early Science and Medicine* 26(4), 289–313. https://doi.org/10.1163/15733823-02630020

Cornelio 1663 = Cornelio, Tommaso. 1663. *Progymnasmata physica*. Venice: Typis Haeredum Franci Baba.

Davies 1954 = Davies, David W. 1954. *The World of the Elseviers 1580–1712*. Dordrecht: Springer.

De Jong 2018 = De Jong, Theo. 2018. *The Most Versatile Scientist, Regent, and VOC Director of the Dutch Golden Age: Johannes Hudde (1628–1704)*. Utrecht: Universiteit Utrecht. MA dissertation.

De Raey 1654 = De Raey, Johannes. 1654. *Clavis philosophiae naturalis, seu Introductio ad naturae contemplationem, Aristotelico-Cartesiana*. Leiden: Ex officina Joannis et Danielis Elsevier.

De Vrijer 1917 = De Vrijer, Marinus Johannes Antoinie. 1917. *Henricus Regius. Een 'cartesiaansch' hoogleraar aan de Utrechtsche hoogeschool*. The Hague: Martinus Nijhoff.

Del Prete 2020 = Del Prete, Antonella. 2020. "Filosofare liberamente a Leida: Adriaan Heereboord, Johannes de Raey, Henricus Bornius." *Dianoia* 31, 57–70.

Denyssen 1682 = Denyssen, Dionysius. 1682. *De heerlyckheyt der Heyligen op Aerde*. Amsterdam: By Hendrick en de weduwe van Dirck Boom.

Des Chene 2001 = Des Chene, Denis. 2001. *Spirits and Clocks, Machine and Organism in Descartes*. Ithaca-London: Cornell University Press.

Descartes 1656 = Descartes, René. 1656. *Principia philosophiae*. Amsterdam: Apud Ludovicum et Danielem Elzevirios.

Descartes 1657–1667 = Descartes, René. 1657–1667. *Lettres de Mr Descartes*, ed. Claude Clerselier. Paris: Chez Charles Angot.

Descartes 1662 = S1 = Descartes, René. 1662. *De homine figuris et Latinitate donatus*, ed. and trans. Florentius Schuyl. Leiden: Apud Petrum Leffen et Franciscum Moyardum.

Descartes 1664a = S2 = Descartes, René. 1664. *De homine figuris et Latinitate donatus*, ed. and trans. Florentius Schuyl. Leiden: Ex officina Hackiana.

Descartes 1664b = C1 = Descartes, René. 1664. *L'homme* (…) *et un Traité de la formation du foetus* (…). *Avec les Remarques de Louis de La Forge*, ed. Claude Clerselier. Paris: Chez Charles Angot.

Descartes 1668–1683 = Descartes, René. 1668–1683. *Epistolae, partim ab auctore Latino sermone conscriptae, partim ex Gallico translatae*. Amsterdam: Apud Danielem Elzevirium; Ex typographia Blaviana.

Descartes 1677a = Descartes, René. 1677. *Tractatus de homine et De formatione foetus, quorum prior Notis perpetuis Ludovici de La Forge* (…) *illustratur*. Amsterdam: Apud Danielem Elsevirium.

Descartes 1677b = C2 = Descartes, René. 1677. *L'homme* (…) *et La formation du foetus, avec les Remarques de Louis de La Forge. A quoy l'on a ajouté Le monde ou Traité de la lumiere* (…). *Seconde edition, reveuë et corrigée*, ed. Claude Clerselier. Paris: Chez Charles Angot.

Descartes 1692 = Descartes, René. 1692. *Proeven der wys-begeerte ofte redenering* (…). *Als mede de verre-gezigtkunde, met een ontwerp van de dierige lichamen, en vorming der vrugt in de baar-moeder* (…). Amsterdam: By Jan ten Hoorn.

AT = Descartes, René. 1974–1986. *Oeuvres. Nouvelle présentation en co-edition avec le Centre national de la recherche scientifique*, ed. Charles Adam and Paul Tannery. Paris: J. Vrin. First edition 1897–1913.

Descartes 1982 = Descartes, René. 1982. *Principles of Philosophy*, ed. and trans. Valentine Rodger Miller and Reese P. Miller. Dordrecht: Springer.

Descartes 2000 = Descartes, René. 2000. *Écrits physiologiques et médicaux*, ed. and intr. Vincent Aucante. Paris: Presses Universitaires de France.

Descartes 2003 = Descartes, René. 2003. *The Correspondence of René Descartes 1643*, ed. Theo Verbeek and Erik-Jan Bos. Utrecht: Zeno, The Leiden-Utrecht Research Institute of Philosophy.

Descartes 2007 = Descartes, René. 2007. *Specimina philosophiae. Introduction and Critical Edition*, ed. Corinna Vermeulen. Utrecht: Zeno, The Leiden-Utrecht Research Institute of Philosophy. PhD dissertation.

Descartes 2011 = Descartes, René. 2011. *De wereld, De mens, Het zoeken naar waarheid*, ed. Erik-Jan Bos and Han van Ruler, trans. Jeanne Holierhoek. Amsterdam: Boom.

Descartes and Regius 2002 = Descartes, René, and Regius, Henricus. 2002. *The Correspondence between Descartes and Henricus Regius*, ed. Erik-Jan Bos. Utrecht: Zeno, The Leiden-Utrecht Research Institute of Philosophy. PhD dissertation.

Di Loreto 1995 = Di Loreto, Mario. 1995. "L'«Inventaire de Stockholm» e il «primo registro» di Descartes. Note in margine alle opere postume sulle matematiche." *Nuncius* 10, 551–615. https://doi.org/10.1163/182539185X00873

Dibon 1990 = Dibon, Paul. 1990. "Clerselier, éditeur de la correspondance de Descartes." in Dibon, Paul. *Regards sur la Hollande du siècle d'or*, 495–522. Naples: Vivarium.

Du Rieu 1875 = Du Rieu, Willem Nikolaas (ed.). 1875. *Album studiosorum Academiae Lugduno-Batavae MDLXXV–MDCCCLXXV. Accedunt nomina Curatorum et Professorum per eadem secula*. The Hague: Apud Martinum Nijhoff.

Fisher 2005 = Fisher, Saul. 2005. "Pierre Gassendi." In *The Stanford Encyclopedia of Philosophy*, ed. Edward N. Zalta. URL = <https://plato.stanford.edu/archives/spr2014/entries/gassendi/>. Accessed 16 August 2022.

Gassendi 1644 = Gassendi, Pierre. 1644. *Disquisitio metaphysica seu Dubitationes et Instantiae adversus Renati Descartes Metaphysicam et Responsa*. Amsterdam: Apud Johannem Blaeu.

Gassendi 1658 = Gassendi, Pierre. 1658. *Opera omnia in sex tomos divisa*. Lyon: Sumptibus Laurentii Anisson, et Joan. Bapt. Devenet.

Gatterer 1785–1792 = Gatterer, Christoph Wilhelm Jacob. 1785–1792. *Anleitung den Harz und andere Bergwerke mit Nuzen zu bereisen*. Göttingen: Im Vandenhöck-Ruprechtischen Verlage.

Hanotaux et al. 1884–1969 = Hanotaux, Gabriel, et al. (eds.). 1884–1969. *Recueil des instructions données aux ambassadeurs et ministres de France depuis les traités de Westphalie jusqu'à la révolution française*. Paris: Félix Alcan et al.

Helk 1971 = Helk, Vello. 1971. *Dänische Romreisen von der Reformation bis zum Absolutismus (1536-1660)*. Copenhagen: Analecta Romana Instituti Danici.

Hobbes 1994 = Hobbes, Thomas. 1994. *The Correspondence of Thomas Hobbes*, ed. Noel Malcolm. Oxford: Clarendon Press.

Huyberts 1652 = Huyberts, Aernouts. 1652. *Disputatio medica inauguralis de affectibus ventriculi circa concoctionem ciborum et appetitus*. Leiden: Apud Franciscum Moyaert. Naples, Biblioteca Nazionale, signature A-B^4

Huygens 1888–1950 = Huygens, Christiaan. 1888–1950. *Œuvres complètes publiées par la Société hollandaise des sciences*, ed. Johan Adriaan Vollgraff. The Hague: Martinus Nijhoff.

Huygens 1892–1899 = Huygens, Constantijn. 1892–1899. *De gedichten van Constantijn Huygens*, ed. Jacob Adolf Worp. The Hague: M. Nijhoff.

Huygens 1911–1917 = Huygens, Constantijn. 1911–1917. *De briefwisseling van Constantijn Huygens, 1608–1697*, ed. Jacob Adolf Worp et al. The Hague: M. Nijhoff.

Jacobson and Hildebrand 1945 = Jacobson, G., and Hildebrand, Bengt. 1945. "Christoff Delphicus Dohna." In *Svenskt Biografiskt Lexikon*, volume 11, 328–333. Stockholm: Albert Bonnier.

Knod 1897 = Knod, Gustav C. 1897. *Urkunden und Akten der Stadt Strassburg. Dritte Abtheilung. Die alten Matrikeln der Universität Strassburg*. Strasbourg: Verlag von Karl J. Trübner.

La Forge 1974 = La Forge, Louis de. 1974. *Oeuvres philosophiques avec une étude bio-bibliographique*, ed. and intr. Pierre Claire. Paris: Presses Universitaires de France.

Lanfrey 1879 = Lanfrey, Pierre. 1879. *L'Église et les philosophes au XVIIIe siècle*. Paris: G. Charpentier. First edition 1857.

Le Clerc 1709 = Le Clerc, Jean. 1709. "Éloge de feu Mr. De Volder Professeur en Philosophie et aux Mathematiques, dans l'Academie de Leide." *Bibliothèque choisie* 18, 346–401.

Lipstorp 1653 = Lipstorp, Daniel. 1653. *Specimina philosophiae Cartesianae. Quibus accedit eiusdem authoris Copernicus redivivus*. Leiden: Apud Johannem et Danielem Elzevier.

Locke 1976–1989 = Locke, John. 1976–1989. *The Correspondence of John Locke*, ed. Esmond Samuel De Beer. Oxford: Clarendon Press.

Matton 2005 = Matton, Sylvain. 2005. "Un témoignage oublié sur le manuscrit du *Traité de l'homme* de Descartes." *Bulletin cartésien* XXXVI/*Archives de Philosophie* 68(1), 19–22.

Mersenne 1933–1988 = Mersenne, Marin. 1933–1988. *Correspondance du P. Marin Mersenne, Religieux Minime*, ed. Cornelius de Waard et al. Paris: Éditions du Centre national de la Recherche Scientifique.

Meschini 2011 = Meschini, Franco A. 2011. "Filologia e scienza. Note per un'edizione critica de *l'Homme* di Descartes." In *Le opere dei filosofi e degli scienziati. Filosofia e scienza tra testo, libro e biblioteca*, ed. Franco A. Meschini and Francesca Puccini, 165–204. Florence: L. S. Olschki.

Meschini 2015 = Meschini, Franco A. 2015. "Per un'edizione critica de *L'Homme* di Descartes. Nuovi materiali ed altre suggestioni." In *L'utopia: alla ricerca del senso della storia, scritti in onore di Cosimo Quarta*, ed. Giuseppe Schiavone, 527–550. Milan: Mimesis.

Meschini 2016 = Meschini, Franco A. 2016. "New Indications for Critical Edition of *L'Homme*." in *Descartes' Treatise on Man and its Reception*, ed. Delphine Antoine-Mahut and Stephen Gaukroger, 49–62. Dordrecht: Springer. https://doi.org/10.1007/978-3-319-46989-8_3

Meyer 1679 = Meyer, Martin, 1679. *Diarii Europaei Continuatio XXXVII*. Frankfurt am Main: Bey Wilhelm Serlins.

Molhuysen 1913–1924 = Molhuysen, Philip Christiaan (ed.). 1913–1924. *Bronnen tot de Geschiednis der Leidsche Universiteit 1574[-1811]*. The Hague: Martinus Nijhoff.

Monchamp 1886 = Monchamp, Georges. 1886. *Histoire du Cartésianisme en Belgique*. Brussels: F. Hayez.

Mourits 2016 = Mourits, Esther. 2016. *Een kamer gevuld met de mooiste boeken. De bibliotheek van Johannes Thysius (1622-1653)*. Leiden: Universiteit Leiden. PhD dissertation.

Muller 2023 = Muller, Jil. 2023. "Humors, Passions, and Consciousness in Descartes's Physiology: The Reconsideration through the Correspondence with Elisabeth." in *Reading Descartes*.

Consciousness, Body, and Reasoning, ed. Andrea Strazzoni and Marco Sgarbi, 59–80. Florence: Firenze University Press. https://doi.org/10.36253/979-12-215-0169-8.05

Nachtomy and Smith 2014 = Nachtomy, Ohad, and Smith, Justin (eds.). 2014. *The Life Science in Early Modern Philosophy*. Oxford: Oxford University Press. https://doi.org/10.1093/acprof:oso/9780199987313.001.0001

Neumann 1935 = Neumann, Alfred. 1935. *The Life of Christina of Sweden*. London: Hutchinson.

Plempius 1653 = Plempius, Vopiscus Fortunatus. 1653. *Fundamenta medicinae ad scholae acribologiam aptata. Editio tertia*. Leuven: Typis ac sumtibus Hiernonymi Nempaei. First edition 1638.

Poelhekke 1961 = Poelhekke, Jan Joseph. 1961. "Nederlandse leden van de *inclyta natio Germanica artistarum* te Padua 1553-1700." *Mededelingen van het Nederlands Historisch Instituut te Rome* 31, 265–373.

Porzio 1704 = Porzio, Lucantonio. 1704. *De motu corporum nonnulla et de nonnullis fontibus naturalibus*. Naples: Impensis Bernardini Gessari.

Prins 1936 = Prins, Izak. 1936. "Gegevens betreffende de 'Oprechte Hollandsche Civet'." *Economisch-Historisch Jaarboek* 20, 3–211.

Ragland 2022 = Ragland, Evan. 2022. *Making Physicians. Tradition, Teaching, and Trials at Leiden University, 1575–1639. Volume 1*. Leiden: Brill. https://doi.org/10.1163/9789004515727

Raymond 1999 = Raymond, Jean-François de. 1999. *Pierre Chanut, ami de Descartes: un diplomate philosophe*. Paris: Beauchesne.

Regius 1641–1643 = Regius, Henricus. 1641–1643. *Physiologia sive Cognitio sanitatis, tribus disputationibus in Academia Ultraiectina publice proposita*. Utrecht: Ex officina Aegidii Roman et al.

Regius 1646 = Regius, Henricus. 1646. *Fundamenta physices*. Amsterdam: Apud Lodovicum Elzevirium.

Revius 1654 = Revius, Jacobus. 1654. *Kartēsiomania, hoc est, Furiosum nugamentum, quod Tobias Andreae, sub titulo Assertionis methodi cartesianae, orbi literato obtrusit, succincte ac solide confutatum*. Leiden: Apud Hieronymum de Vogel.

Ritter 1705 = Ritter, L. A. 1705. *De principum in Cartesium favore*. Greifswald: Literis G. H. Adolphi, 1705.

Schmaltz 2016 = Schmaltz, Tad M. 2016. "The Early Dutch Reception of *L'Homme*." In *Descartes' Treatise on Man and Its Reception*, ed. Delphine Antoine-Mahut and Stephen Gaukroger, 71–90. Dordrecht: Springer. https://doi.org/10.1007/978-3-319-46989-8_5

Schneider 1655 = Schneider, Konrad Victor. 1655. *Liber de osse cribriformi, et sensu ac organo odoratus, et morbis ad utrumque spectantibus*. Wittenberg: Typis Jobi Wilhelmi Fincelij.

Sennert 1628 = Sennert, Daniel. 1628. *Institutionum medicinae libri V*. Wittenberg: Apud haeredes Zach. Schüreri Sen. First edition 1611.

Sorbière 1660 = Sorbière, Samuel. 1660. *Lettres et discours (…) sur diverses matières curieuses, ont été imprimées*. Paris: Chez François Clousier.

Sorbière 1691 = Sorbière, Samuel. 1691. *Sorberiana sive Excerpta ex ore Samuelis Sorbiere*. Toulouse: Typis Guill. Lud. Colomyez, Hier. Posuël, M. Fouchac et G. Bely, bibliopolas.

Steensen 1667 = Steensen, Niels. *Elementorum myologiae specimen*. Florence: Ex typographia sub signo stellae.

Stock and Weichert 2020 = Stock, Jürgen, and Weichert, Reiner. 2020. *Die Hartzings. Der Aufstieg einer Moerser Familie unter der Ostindischen Kompanie*. Duisburg: Mercator.

Strazzoni 2014 = Strazzoni, Andrea. 2014. "On Three Unpublished Letters of Johannes de Raey to Johannes Clauberg." *Noctua* 1(1), 66–103. https://doi.org/10.14640/NoctuaI3

Strazzoni 2020 = Strazzoni, Andrea. 2020. "A Letter of Peter Hartzing to Gerhard Wolter Molanus." *Noctua* 7(1), 158–181. https://doi.org/10.14640/NoctuaVII4

Strazzoni 2022 = Strazzoni, Andrea. 2022. "Some Unpublished Fragments on Descartes's Life and Works." *The Seventeenth Century* 37(5), 801–839. https://doi.org/10.1080/0268117X.2021.2021976

Strazzoni 2023a = Strazzoni, Andrea. 2023. "The Use and Plagiarism of Descartes's *Traité de l'homme* by Henricus Regius: A Reassessment." *Perspectives on Science* 31(5), 627–683. https://doi.org/10.1162/posc_a_00587

Strazzoni 2023b = Strazzoni, Andrea. 2023. "Neglected Sources on Cartesianism: The Academic *Dictata* of Johannes de Raey." *Intellectual History Review* 33(4), 525–586. https://doi.org/10.1080/17496977.2022.2038466

Swammerdam 1975 = Swammerdam, Johannes. 1975. *The Letters of Jan Swammerdam to Melchisédech Thévenot: With English Translation and a Biographical Sketch*, ed. and trans. Gerrit Arie Lindeboom. Amsterdam: Taylor & Francis.

Tessin 1965 = Tessin, Georg. 1965. *Die deutschen Regimenter der Krone Schweden*. Köln-Graz: Böhlau Verlag.

Thijssen-Schoute 1967 = Thijssen-Schoute, Caroline Louise. 1967. *Uit de republiek der letteren: 11 studiën op het gebied de ideeëngeschiedenis van den Gouden Eeuw*. The Hague: Martinus Nijhoff.

Ulfeldt 1949 = Ulfeldt, Leonora Christina. 1949. *Jammers minde og andre selvbiografiske skildringer*. Copenhagen: Rosenkilde og Bagger.

Van Anrooij and Hoftijzer 2017 = Van Anrooij, Wim, and Hoftijzer, Paul. 2017. *Vijftien strekkende meter: nieuwe onderzoeksmogelijkheden in het archief van de Bibliotheca Thysiana*. Hilversum: Verloren.

Van Hogelande 1646 = Van Hogelande, Cornelis. 1646. *Cogitationes, quibus Dei existentia, item animae spiritalitas, et possibilis cum corpore unio, demonstrantur, nec non, brevis historia oeconomiae corporis animalis, proponitur, atque mechanice explicatur*. Amsterdam: Apud Ludovicum Elzevirium.

Van Otegem 2002 = Van Otegem, Matthijs. 2002. *A Bibliography of the Works of Descartes (1637–1704)*. Utrecht: Zeno, The Leiden-Utrecht Research Institute of Philosophy. PhD dissertation.

Van Roijen et al. 2013 = Van Roijen, R., De Jonge, H., and Wieles, W. B. 2013. *Inventaris van de archieven van de Bibliotheca Thysiana en van leden van de familie Thijs en aanverwante families, 16de-20ste eeuw (ATH). Versie 19 november 2013*. Leiden. URL = <http://hdl.handle.net/1887.1/item:1918704>. Accessed 29 August 2022.

Verbeek 2003 = Verbeek, Theo. 2003. "Heereboord, Adriaan (1614–1661)." In *Dictionary of Seventeenth and Eighteenth-Century Dutch Philosophers*, ed. Wiep van Bunge, Henri Krop, Han van Ruler, and Paul Schuurman, volume 1, 395–397. Bristol: Thoemmes.

Wackernagel 1962 = Wackernagel, Hans Georg. 1962. *Die Matrikel der Universität Basel. Band III: 1601/02–1665/66*. Basel: Böhlau Verlag.

Westera 2018 = Westera, Lambert Douwe. 2018. *'Met list en vlijt'. Koningen & kooplieden en de kanonnenhandel tijdens de Republiek*. Amsterdam: Universiteit van Amsterdam. PhD dissertation.

Wittich 1653 = Wittich, Christoph. 1653. *Dissertationes duae, quarum Prior de S. Scripturae in rebus philosophicis abusus examinat (…); altera dispositionem et ordinem totius universi (…) tradit*. Amsterdam: Apud Ludovicum Elzevirium.

Wrangel 1891 = Wrangel, Fredrik Ulrik. 1691. *Liste des diplomates français en Suède 1541–1891*. Stockholm: Imprimerie Royale-P. A. Norstedt & Söner.

Index of Names

A
Agostini, Siegrid, 30
Albinus, Frederick Bernhard, 39
Alexandrescu, Vlad, 11
Andreae, Tobias, 2, 4, 5, 28–30
Antoine-Mahut, Delphine, 2
Aristotle, 17
Aucante, Vincent, 1
Auzout, Adrien, 23
Avicenna, 42

B
Baillet, Adrien, 29
Bartholin, Rasmus, 23
Bartholin, Thomas, 25
Belgioioso, Giulia, 26
Bitbol-Hespériès, Annie, 26
Boissière, Gustave, 14
Borch, Ole, 17
Bornius, Henricus, 11
Bos, Erik-Jan, 2
Breyne, Jakob, 23
Burman, Frans, 31

C
Caps, Géraldine, 1
Chanut, Pierre-Hector, 8
Chevreau, Urbain, 8
Christina of Sweden, 14
Clauberg, Johannes, 29
Clerselier, Claude, 1–3

Colvius, Andreas, 3, 10
Cook, Harold J., 31
Cornelio, Tommaso, 23

D
Davies, David W, 22
De Bils, Lodewijk, 25
De Geer, Lodewijk, 29
De Jong, Theo, 25
De Jonge, H., 2, 39
De Raey, Elisabeth, 1–5, 14–22, 24, 25, 33, 36, 39, 42, 45
De Raey, Johannes, 1–3, 5, 12, 23, 37, 39
De Volder, Burchard, 24
De Vrijer, Marinus Johannes Antoinie, 12
De Witt, Johan, 31
Del Prete, Antonella, 14
Denyssen, Dionysius, 24
Des Chene, Denis, 1
Descartes, René, 1, 3
Di Loreto, Mario, 36
Dibon, Paul, 29
Dohna, Christopher Delphicus von, 14
Du Rieu, Willem Nikolaas, 24

E
Elisabeth of Bohemia, 1–3, 45
Elzevier, Daniel, 4, 16, 22, 33, 35, 37
Elzevier, Louis, 12, 37

F
Fernel, Jean, 41, 43
Fisher, Saul, 11
Frederik Hendrik (Stadtholder), 13
Furly, Benjamin, 24

G
Galen, 42
Gassendi, Pierre, 11
Gatterer, Christoph Wilhelm Jacob, 24
Gaukroger, Stephen, 2
Graevius, Johann Georg, 24
Guenellon, Pieter, 24

H
Hanotaux, Gabriel, 28
Hartzing, Peter, 23, 25
Heereboord, Adriaan, 1–3, 8
Helk, Vello, 24
Hildebrand, Bengt, 14
Hobbes, Thomas, 12
Hoftijzer, Paul, 39
Hudde, Johannes, 24, 25
Huyberts, Aernout, 1, 4, 5, 20, 22, 23, 25
Huygens, Christiaan, 29, 35
Huygens, Constantijn, 3, 9, 10, 12, 31

J
Jacobson, G, 13, 14
Johann Friedrich, Duke of Brunswick-Lüneburg, 23
Johnson, Samson, 3, 5

K
Knod, Gustav C, 24
Konrad Victor Schneider, 41

L
La Forge, Louis de, 4, 19
La Voyette, Louis de, 3, 9, 13
Lanfrey, Pierre, 14
Langlij, Lijdia, 23
Le Clerc, Jean, 24
le Jeûne, Guillaume *see* Wetstein, Henricus, 38
Legrand, Jean-Baptiste, 32
Lipstorp, Daniel, 35
Locke, John, 24

M
Maire, Johannes, 37
Matton, Sylvain, 8
Melder, Christiaan, 23
Mersenne, Marin, 6
Meschini, Franco A, 2
Meyer, Martin, 28
Molhuysen, Philip Christiaan, 24
Monchamp, Georges, 19
More, Henry, 30
Mourits, Esther, 39
Muller, Jil, 13

N
Nachtomy, Ohad, 1
Neumann, Alfred, 36
Nissel, Johann Georg, 17

P
Philippi, Guillaume, 4, 39
Plempius, Vopiscus Fortunatus, 36
Poelhekke, Jan Joseph, 24
Pollot, Alphonse, 1, 3, 7
Porzio, Lucantonio, 23
Prins, Izak, 24

R
Ragland, Evan, 1
Raymond, Jean-François de, 28
Regius, Henricus, 2, 3, 10
Revius, Jacobus, 32
Ritter, L. A, 16

S
Sallo, Denis de, 26
Schalbruch, Johann Theodor, 18
Schmaltz, Tad M, 10
Schneider, Konrad Victor, 41
Schoock, Martin, 30
Schuyl, Florentius, 1, 3
Sennert, Daniel, 16
Settala, Ludovico, 42
Sladus, Matthaeus, 24
Smith, Justin, 1
Sorbière, Samuel, 11
Steensen, Niels, 23
Stock, Jürgen, 24
Strazzoni, Andrea, 1, 47
Stuart, Elizabeth, 11
Swammerdam, Johannes, 23

Index of Names

T
Tassius, Johann Adolf, 35
Tessin, Georg, 13
Thévenot, Melchisédech, 24
Thijssen-Schoute, Caroline Louise, 31
Thysius, Johannes, 39
Tiele, P. A., 39

U
Ulfeldt, Leonora Christina, 13

V
Van Anrooij, Wim, 39
Van Buitendijck, Petrus, 32
Van Gutschoven, Gerard, 4
Van Hogelande, Cornelis, 2
Van Otegem, Matthijs, 2
Van Roijen, R, 2, 39
Van Schooten, Frans jr, 33
Veen, Egbertus, 24
Verbeek, Theo, 15
Vesalio, Andrea, 41
Vorstius, Adolph, 14, 16, 23, 24

W
Wackernagel, Hans Georg, 24
Weichert, Reiner, 24
Westera, Lambert Douwe, 24
Wetstein, Henricus, 38
Wieles, W. B., 2, 39
Wittich, Christoph, 32
Wrangel, Fredrik Ulrik, 28

Index of Manuscripts

A
Amsterdam, Stadsarchief
  toegangsnummer 5001,
    inventarisnummer 41, 24
  toegangsnummer 728,
    inventarisnummer 37, 24
Amsterdam, Universiteitsbibliotheek
  ms. K 137, 12

C
Copenhagen, Det Kongelige Bibliotek
  ms. Don. var. nr. 145 4to

G
Groningen, Universiteitsbibliotheek
  ms. GN108, 33

H
Hamburg, Staats- und
  Universitätsbibliothek Carl von
  Ossietzky
  Cod. phil. 323 W. 28

L
Leiden, Erfgoed Leiden en Omstreken
  toegangsnummer 1004,
    inventarisnummer 222, 25
  toegangsnummer 1004,
    inventarisnummer 23 Y, 24
Leiden, Universiteitsbibliotheek
  ms. ATH 1444, 1
London, Wellcome Library
  ms. 3415

P
Paris, Bibliothèque Nationale de France
  MS latin 10352

T
The Hague, Koninklijke Bibliotheek
  ms. KA 45, 13

U
Uppsala, Universitetsbiblioteket
  Waller Ms benl-00586
Utrecht, Het Utrechts Archief
  toegangsnummer 34-4,
    inventarisnummer U048a004, 24
  toegangsnummer 711,
    inventarisnummer 129, 24
  toegangsnummer 711,
    inventarisnummer 99, 24

MIX
Papier aus verantwortungsvollen Quellen
Paper from responsible sources
FSC® C105338

If you have any concerns about our products,
you can contact us on
ProductSafety@springernature.com

In case Publisher is established outside the EU,
the EU authorized representative is:
**Springer Nature Customer Service Center GmbH
Europaplatz 3, 69115 Heidelberg, Germany**

Printed by Libri Plureos GmbH
in Hamburg, Germany